NASA AERON...   ...ES

Crash
Course

...d from
Accidents in...ving
Remotely Piloted and
Autonomous Aircraft

Peter W. Merlin

Library of Congress Cataloging-in-Publication Data

Merlin, Peter W., 1964-
  Crash course : lessons learned from accidents involving remotely piloted and
autonomous aircraft / Peter W. Merlin.
      pages cm. -- (NASA SP ; 2013-600)
  Includes bibliographical references and index.

1. Drone aircraft--United States--Testing. 2. Research aircraft--Accidents--
United States. 3. Flight testing. 4. NASA Dryden Flight Research Center. I.
United States. National Aeronautics and Space Administration. Aeronautics
Research Mission Directorate. II. Title.
  UG1242.D7M47 2013
  629.133'34--dc23
                                   2012046389

ISBN 978-1-62683-000-4
90000>

9 781626 830004

Table of Contents

Acknowledgments

The author would like to thank the many people who helped us make this book possible. First of all, thanks to Tony Springer, NASA Aeronautics Research Mission Directorate, for sponsoring this project. I am grateful for the efforts of many people at NASA Dryden Flight Research Center including, but not limited to, Dr. Christian Gelzer, Tom Tschida, and Karl Bender. Thanks to Chris Yates and Ben Weinstein at Media Fusion for preparing the manuscript for publication. Special thanks to Victoria Regenie, Mark Pestana, Ken Cross, and Patrick Stoliker—who reviewed the material for technical accuracy—and, especially, to Sarah Merlin for copyediting the final manuscript.

Divergent roll oscillatory motion caused by excessive control-system gain resulted from inadequate analytical modeling that allowed engineers to overestimate system safety margins. NASA

Introduction

For over half a century, engineers and scientists have worked to develop remotely piloted and autonomous aircraft. Evolving from simple radio-controlled models to sophisticated aircraft equipped with fly-by-wire controls, advanced composite structures, and integrated propulsion, some of these aircraft require complex ground stations staffed by pilots/operators, systems monitors, engineers, and scientists. As a result, they eventually came to be considered unmanned aircraft systems (UASes).

A wide variety of UASes have been developed for civil and military applications. Dedicated experimental models known as remotely piloted research vehicles (RPRVs) complement piloted experimental aircraft; provide affordable "quick look" design validation; enable hazardous testing without risk to pilots' lives; and offer new capabilities such as high-altitude, solar-powered environmental monitoring or advanced propulsion systems. Operational UASes have been used for law enforcement, firefighting, science, and agriculture. Military applications include reconnaissance, strike, communications, and cargo delivery.

Terminology used to describe these craft is constantly evolving. The term "unmanned" was replaced with "unpiloted" or "uninhabited," but both were ultimately considered inadequate. In the popular press, the word "drone" is frequently used to describe any type of aircraft that does not accommodate an onboard crew. Unfortunately, this gives the impression that the aircraft is simply a mindless robot. Nothing could be further from the truth; there are always numerous people involved, serving as operators, monitors, and ground crew.

The terms "remotely piloted" and "autonomous" have been used to distinguish between aircraft controlled by an operator using conventional stick-and-rudder-type controls in a ground cockpit (human-in-the-loop) and those fully controlled by an onboard computer that receives navigation input from an operator on the ground (human-on-the-loop). In 2010, U.S. Air Force leadership elected to introduce an all-encompassing new term—remotely piloted aircraft, or RPA—in order to recognize that in every case, there are humans involved regardless of the level of autonomy. Air Force officials wanted to impress upon the Federal Aviation Administration (FAA) and the aerospace community that these vehicles are always under positive control despite the lack of a human crew on board.[1] Whatever they are called, RPA are seeing increasing use as military, commercial, and research tools.

Between 1943 and 1959, prior to the advent of practical RPRVs, more than two dozen piloted research airplanes and prototypes were lost in accidents, more than half of which were fatal. By the 1960s, researchers began to recognize the value of using remotely piloted vehicles as one means of mitigating risks associated with flight testing.[2] Remotely piloted vehicles had been developed for military uses as early as World War I, but these drone weapon systems were technologically crude and designed simply to fly a one-way mission from which the vehicle would not return. Advances in electronics during the 1950s greatly increased the reliability of control systems, rendering development of RPRVs more practical. Early efforts focused on guidance and navigation, stabilization, and remote control. Eventually, designers worked to improve technologies to support these capabilities through integration of improved avionics, microprocessors, and computers. The RPRV concept was attractive to researchers because it built confidence in new technology through demonstration under actual flight conditions, at relatively low cost, in quick response to demand, and at no risk to the pilot.

Though the use of remotely piloted vehicles in place of aircraft requiring onboard human crews offers advantages, there are significant tradeoffs. Taking the pilot out of the airplane can sometimes mean savings in terms of development and fabrication. The cost and complexity of some robotic and remotely piloted vehicles are less than those of comparable aircraft requiring an onboard crew, since there is no need for life-support systems, escape and survival equipment, or hygiene facilities. On the other hand, costs for ground-support equipment and personnel may increase along with requirements for complex ground control stations. Additionally, with vehicles lacking onboard crews, hardware costs may decrease while software costs increase. There is, however, definite benefit in terms of aircrew safety. Hazardous testing that would pose unacceptable risk to an onboard pilot can be undertaken with a vehicle considered expendable or semiexpendable.

Quick response to customer requirements and reduced program costs resulted from the elimination of redundant systems (usually added for crew safety) and man-rating tests and through the use of less-complex structures

---

1. "A drone by any other name…," Reuters Blog, Dec. 14, 2009, *http://blogs.reuters.com/ summits/2009/12/14/a-drone-by-any-other-name/*, accessed April 30, 2010.
2. R. Dale Reed, "Flight Research Techniques Utilizing Remotely Piloted Research Vehicles," Technical Report (TR) AGARD-LS-108, Paper No. 8, Research Engineering, NASA Dryden Flight Research Center (DFRC), Edwards, CA (1980).

and systems. Subscale test vehicles may cost less than full-size airplanes while providing usable aerodynamic and systems data. The use of programmable ground-based control systems provides additional flexibility and eliminates downtime resulting from the need for extensive aircraft modifications.[3]

Researchers at the National Aeronautics and Space Administration's (NASA) Dryden Flight Research Center (DFRC), Edwards, CA, have used RPRVs to pioneer innovative new concepts and contribute to the development of technologies for current and future aircraft. Low-budget radio-controlled models, as well as more complex subscale and full-scale research aircraft, have been successfully used for tasks that would have been hazardous or physically challenging for a human crew. Experimental and operational use of this cutting-edge technology has also resulted in a number of mishaps that may provide valuable lessons for future RPRV and UAS operators, and indeed for anyone involved in aircraft operations.

The following investigation of RPRV/UAS mishaps will examine their causes, consequences, resultant corrective actions, and lessons learned. Most undesired outcomes usually do not occur because of a single event, but rather from a series of events and actions involving equipment malfunctions and/or human factors. This book comprises a series of case studies focusing mostly on accidents and incidents involving experimental aircraft. The information provided should be of use to flight-test organizations, aircraft operators, educators, and students, among others. These lessons are not unique to the UAS environment and are also applicable to human aviation and space flight activities. Common elements include crew resource management, training, mission planning issues, management and programmatic pressures (e.g., schedule, budget, resources), cockpit/control station design, and other factors.

---

3. Ibid.

Use of a three-eighths-scale remotely piloted model of the F-15, seen here beside its full-size counterpart, allowed researchers to conduct hazardous tests without risking the pilot's life. NASA

CHAPTER 1

F-15 RPRV

The First Practical Remotely Piloted Research Vehicle

NASA researchers have been experimenting with remotely piloted aircraft since the early 1960s, when radio-controlled models were used to test the aerodynamics of wingless space-reentry vehicle configurations. As the merits of this technique became apparent, engineers quickly began designing more sophisticated test beds. The Agency's involvement with the remotely piloted vehicle concept came of age a decade later, when researchers applied this new technology to support research, development, testing, and evaluation of a new Air Force fighter, the McDonnell-Douglas F-15 Eagle.

A Low-Cost, Low-Risk Alternative

In 1969, the Air Force selected McDonnell-Douglas Aircraft Corporation to build the F-15, a Mach 2–capable air superiority fighter airplane designed using lessons learned during aerial combat over Vietnam. The prototype first flew in July 1972. In the months leading up to that flight, Maj. Gen. Benjamin Bellis, chief of the F-15 System Program Office at Wright-Patterson Air Force Base, OH, requested NASA assistance in testing a three-eighths-scale model F-15 RPRV to explore aerodynamic and control-system characteristics of the F-15 configuration in spins and high-angle-of-attack flight. Such maneuvers can be extremely hazardous. Rather than risk harm to a valuable test pilot and prototype, a ground pilot would develop stall-/spin-recovery techniques with the RPRV and pass lessons learned on to test pilots flying the actual airplanes.

In April 1972, NASA awarded McDonnell-Douglas a $762,000 contract to build three F-15 RPRV models. Other contractors provided electronic components and parachute recovery equipment. NASA technicians installed avionics, hydraulics, and other subsystems. The F-15 RPRV was 23.5 feet long, made primarily of fiberglass and wood, and weighed 2,500 pounds. It had no propulsion system and was designed for midair recovery using a helicopter.

The F-15 RPRV ground cockpit contained a typical array of flight instruments. A forward-pointing television camera in the aircraft provided an outside visual reference. NASA

Each model cost just over $250,000, compared to $6.8 million for a full-scale F-15 aircraft.[1]

Every effort was made to use off-the-shelf components and equipment readily available at the Flight Research Center, including hydraulic components, gyros, and telemetry systems from the lifting body research programs. An uplink, then being used for instrument-landing-system experiments, was acquired for the RPRV ground control station. The ground cockpit was fashioned from a general-purpose simulator that had been used for stability-and-control studies. Data-processing computers were adapted for use in a programmable ground-based control system. A television camera provided forward visibility. At the end of each test flight, the RPRV was to be plucked from the sky using a helicopter equipped with a midair recovery system (MARS). This parachute

1. Richard P. Hallion and Michael H. Gorn, *On the Frontier: Experimental Flight at NASA Dryden* (Washington, DC: Smithsonian Books, 2003), p. 210.

mechanism had been successfully used on the Ryan AQM-34 Firebee drone, but it was to prove troublesome during the F-15 RPRV test program.[2]

The first F-15 RPRV arrived at the Flight Research Center in December 1972 but was not flown until October 12, 1973. The model was carried to an altitude of about 45,000 feet beneath the wing of a modified B-52 Stratofortress. Following release from the launch pylon at a speed of 175 knots, ground pilot Einar Enevoldson guided the craft through a flawless nine-minute flight during which he explored the vehicle's basic handling qualities. At 15,000 feet altitude, a 12-foot spin-recovery parachute deployed to stabilize the descent. An 18-foot engagement chute and a 79-foot-diameter main chute then deployed so that the RPRV could be snagged in flight by a hook and cable beneath a helicopter and set down gently on an inflated bag.[3]

Flying the RPRV

For pilots accustomed to flying conventional airplanes, the experience of pilot-ing the RPRV took some getting used to. Einar Enevoldson considered the task of flying the RPRV both physically and psychologically challenging. The lack of physical cues left him feeling detached from the essential reassuring sensations of flight that ordinarily provide a pilot with situational feedback. Lacking sensory input, he found that his workload increased and subjective time seemed to speed up. Afterward, he reenacted the mission in a simulator at 1.5 times actual time and found that the pace seemed the same as it had during the flight.[4]

Researchers had monitored his heart rate during the flight to see if it would register the 70 to 80 beats per minute typical for a piloted test flight. They were surprised to see the readings indicate 130 to 140 beats per minute as the pilot's stress level increased. Enevoldson considered flying the F-15 RPRV less pleasant and satisfying than he normally found flying any other difficult or demanding test mission.

"The results were gratifying," he wrote in his postflight report, "and some satisfaction is gained from the success of the technical and organizational achievement—but it wasn't fun."[5]

---

2. Reed, "Flight Research Techniques Utilizing Remotely Piloted Research Vehicles."

3. Hallion and Gorn, *On the Frontier*, pp. 210–211.

4. F-15 Drone Flight Report, Flight Number D-1-3, Oct. 12, 1973, DFRC Historical Reference Collection, NASA DFRC, Edwards, CA, file location L3-8-4A-1.

5. Ibid.

The unpowered RPRV was carried to altitude beneath the wing of a modified B-52. After release, the ground pilot performed test maneuvers to develop stall- and spin-recovery techniques. NASA

In subsequent tests, Enevoldson and other research pilots explored the vehicle's stability and control characteristics. Spin testing confirmed the RPRV's capabilities for returning useful data, encouraging officials at the F-15 Joint Test Force to proceed with piloted spin trials in the preproduction prototypes at Edwards.[6]

Bill Dana piloted the fourth F-15 RPRV flight on December 21, 1973. He collected about 100 seconds of data at angles of attack exceeding 30 degrees and with 90 seconds of control-response data. Dana had a little more difficulty controlling the RPRV in flight than he had in the simulator but otherwise felt everything went well. At Enevoldson's suggestion, the simulator flights had been sped up to 1.4 times actual speed, and Dana later acknowledged that this had provided a more realistic experience.[7]

---

6. Hallion and Gorn, *On the Frontier*, p. 211.

7. F-15 Drone Flight Report, Flight Number D-4-6, Dec. 21, 1973, DFRC Historical Reference Collection, NASA DFRC, Edwards, CA, file location L3-8-4A-2.

During a postflight debriefing, Dana was asked how he liked flying the RPRV. He responded that it was quite different from sitting in the cockpit of an actual research vehicle, where he generally worried and fretted until just before launch. Then he could settle down and just fly the airplane. With the RPRV, he said he was calm and cool until launch and then felt keyed up through the recovery.[8]

Midair Recovery

The MARS helicopter recovery crew consisted of a pilot, copilot, flight engineer, and winch operator. The helicopter's aft cargo door had been removed, but the cargo ramp remained in place. A winch, located under the main rotor head, controlled a cable that extended downward through a hole in the bottom of the fuselage and back to a harness between two extended poles that served as the capture device.

As with any flight research program, safety was a major concern. The F-15 RPRV Operational Readiness Review committee examined every aspect of the vehicle and associated systems in order to ensure the safety of the RPRV as well as that of personnel and property on the ground. MARS was thoroughly scrutinized due to the complex nature of both the equipment and the procedures used to snag the RPRV in midair. Additionally, since vehicle weight restrictions precluded installation of redundant subsystems, the parachute was considered the sole backup for preventing loss of the RPRV in the event of malfunction. Due to the vital nature of the recovery system, every effort was made to ensure that it would function properly.[9]

The RPRV was equipped with a system of four parachutes, a release mechanism, pyrotechnic devices, and other equipment. Two small chutes deployed to begin the stabilization and deceleration sequence. Next, a large (80-foot-diameter) main chute slowed the vehicle's descent while a smaller (18-foot-diameter) engagement chute opened about 30 feet above the main chute to provide a target for the capture device. In a MARS recovery, the helicopter pilot had to maneuver above the target vehicle while the winch operator snagged the engagement chute with the cable and reeled the RPRV to within about 20 feet of the helicopter for return to base.

---

8. Ibid.

9. Memorandum from Robert W. Borek, chairman of the F-15 RPRV Ad Hoc Committee, to the project manager, Sept. 7, 1973, DFRC Historical Reference Collection, NASA DFRC, Edwards, CA, file location L1-7-5B-2.

During initial test flights, the crew of a CH-3E helicopter snagged the RPRV in midair and delivered it to a safe landing. NASA

Success of this maneuver required the proper function of a parachute release mechanism, attached with nylon straps between the engagement parachute's riser lines and the RPRV. A link assembly and load line connected the mechanism to the main parachute. During normal operation, when the main-chute load line was pulled upon helicopter engagement, the mechanism enabled the main chute to separate. The RPRV was then left suspended beneath the helicopter on the load line, where it could be safely winched up to its ferry position.[10]

Plans called for the recovery helicopter to circle the drop zone at an altitude of 10,000 feet above the ground. Chute deployment was controlled by a barometric device, which could be adjusted to activate at altitudes between 7,000 and 15,000 feet. Successful midair retrieval required skillful operation of the retrieving aircraft (in this case, a CH-3E helicopter), favorable atmospheric conditions, and successful execution of a difficult maneuver to snag the target vehicle. Due to their maneuverability, helicopters were the optimal aircraft for such operations.

---

10. James R. Phelps, Harry R. Chiles, and William R. Petersen, "Report on F-15 RPRV Recovery Incident of July 10, 1974," Aug. 6, 1974, DFRC Historical Reference Collection, NASA DFRC, Edwards, CA, file location L1-7-5B-2.

Failure to Engage

On July 10, 1974, the first of several incidents involving the MARS parachute gear occurred during the ninth flight. After successfully releasing the RPRV from the B-52, ground pilot Einar Enevoldson performed maneuvers to obtain stability and control data at negative angles of attack. Following completion of all test points, he activated the recovery parachute system.

During recovery, the main parachute failed to separate upon engagement with the helicopter's parachute capture system. The hung chute produced a strong drag force on the helicopter and began pulling more load line from the winch drum. The helicopter crew attempted to force main-chute separation by increasing winch brake friction on the line, which finally broke. At no time did the crew attempt to sever the load line using the emergency cable cutter. Fortunately, the CH-3E sustained no damage.

The F-15 RPRV, hanging beneath the main parachute and trailing 400 feet of load line, landed on the Edwards Precision Impact Range Area (PIRA). Normally, upon impact, a gravitational force (g)–sensing switch located in the fuselage would have activated an explosive bolt to release the main chute, but a faulty microswitch in the parachute release mechanism deactivated this circuit. Before the main chute had a chance to collapse, it was caught by the wind, dragging the RPRV nearly a quarter of a mile across the desert.[11]

An accident investigation board determined that several factors contributed to the mishap. The primary cause was human error on the part of the F-15 RPRV crew chief that resulted in improper assembly of the parachute release mechanism, as well as failure of an inspector to catch the error prior to flight. There were also several contributing factors, the most serious of which was faulty design by the manufacturer that allowed the release mechanism to be incorrectly installed. Documentation of procedures for installing the device appeared adequate but was inefficiently spread over several separate documents, a situation that prevented critical items from coming to light and required a great deal of cross-checking. Lack of recent practical experience on the part of the mechanic and inspector also played a part, and experienced personnel who had performed the installation on prior flights were not available for consultation. The regular inspector was ill and had been replaced by a backup, and the mechanic most familiar with the parachute system had been transferred to a different project. Investigators also noted that the Air Force MARS crew had

---

11. Ibid.

previously experienced an identical incident, but they had not shared this data with F-15 RPRV project personnel.[12]

During the investigation, it became apparent that the same type of error could have easily occurred in other research programs as a result of prevalent practices at the Flight Research Center. Inspectors assigned to many research programs, due to the unique and complex nature of the systems involved, often lacked current practical experience in areas for which they were responsible. Investigators felt that each inspector should be fully knowledgeable, having mastered the practical and theoretical aspects of a job before being able to determine whether others were doing it properly. This objective required that any inspector be part of the design team as early in the program as possible. The board recommended that only the most knowledgeable person should have responsibility for final system verification and that provisions be made to ensure the availability of adequately trained backup personnel.

Investigators found it was also common practice to remove experienced team members from projects without ensuring the availability of qualified replacements or remaining personnel. Consequently, they suggested that shuffling of project members could result in loss of interest and low morale, adversely affecting efficiency and safety.

In order to ensure the successful implementation of research projects, Flight Research Center managers had prescribed procedures for reviewing any program to determine its design, planning, and functional adequacy. However, because of the diversity and complexity of many programs, it was often difficult to staff a review committee. Review personnel were typically drawn from other programs to which they were assigned full time. Because of such commitments, they were unable to devote sufficient time to the review. Investigators felt that, to be effective, review teams should be appointed early in any given program, and they should use a minimum amount of time over the program's buildup stages to become knowledgeable about various facets of the project and establish continuity with the project team.

Finally, investigators noted that utilization of remotely piloted vehicles as research tools was likely to increase in the future. They felt that since various phases of testing, such as launch and recovery, would require the use of conventional piloted aircraft, implementation of a system-safety fault-hazard analysis would be warranted to ensure safe flight operations.[13]

---

12. Ibid.
13. Ibid.

Alternative Landing Options

Aside from the technical difficulty involved in a MARS recovery, there were also schedule considerations. Air Force helicopters were not always available on preferred flight days. After reviewing the rate of flights per month, several project engineers recommended instituting a soft-landing procedure in which parachutes would lower the RPRV gently to the ground. They reasoned that eliminating the helicopter would shorten the interval between tests, allowing RPRV flights to continue as scheduled, dependent only on weather and availability of the B-52. The soft-landing option—known as the range recovery system, or RRS—would serve only as a backup, however, with MARS as the primary recovery technique whenever the helicopter was available.[14]

Program managers examined several factors before approving the new landing method. First, wind conditions at the landing site had the potential to damage the vehicle. Despite a consensus that winds in excess of 5 knots would likely drag the RPRV, reducing the maximum windspeed limit below 10 knots was considered impractical due to prevailing local weather conditions. A maximum windspeed of 10 knots was accepted because imposing a stricter limit would have simply substituted one potential source of delay, weather constraint, for another, helicopter availability.[15]

To reduce the risk of dragging, a ground-proximity "whisker switch" was installed to activate a parachute release mechanism upon touchdown. A type of switch used as standard equipment on the Firebee drone was recommended.[16] The switch was deemed to be of sufficient electrical and mechanical integrity, and sequencing of the circuitry in which it was installed adequately allowed for safe parachute release.[17]

Alternating between MARS and RRS techniques from flight to flight as conditions warranted introduced the opportunity for human error. The recovery system required different programming input prior to flight, depending on which technique was to be used. At the direction of the F-15 RPRV Operational Readiness Review Ad Hoc Committee chairman, foolproof procedures were written to ensure proper programming of the recovery system.

---

14. Minutes of the F-15 RPV ORR Committee Meeting, Sept. 30, 1974, DFRC Historical Reference Collection, NASA DFRC, Edwards, CA, file location L1-7-5B-2.
15. Minutes of the F-15 RPV ORR Committee Meeting, Oct. 2, 1974, DFRC Historical Reference Collection, NASA DFRC, Edwards, CA, file location L1-7-5B-2.
16. Ibid.
17. Minutes of the F-15 RPV ORR Committee Meeting, Oct. 4, 1974, DFRC Historical Reference Collection, NASA DFRC, Edwards, CA, file location L1-7-5B-2.

Engineers wrote two documents defining MARS/RRS electrical connection and test procedures and F-15 RPRV configuration changeover procedures, verifying them through functional tests with the vehicle.[18]

A section of the PIRA was inspected and deemed satisfactory for safe landing operations. Simulations validated that the new landing procedures would ensure recovery within the prescribed region. Subsequently, the F-15 RPRV Ad Hoc Committee gave the project manager approval to use the RRS soft-landing procedure whenever necessary. The committee chairman noted, however, that "although the committee approved project go-ahead, the physical survivability of the F-15 RPRV [was] not a committee responsibility."[19]

Link Mechanism Failure

As it turned out, the helicopter was available for the next two scheduled flights. Successful MARS recoveries were made in both cases. Trouble struck again on October 16, 1974, during the 14th flight.

Approximately 25 seconds after initiation of the RPRV's recovery sequence, the main chute separated and collapsed, leaving the vehicle to descend using only the 18-foot engagement chute. The RPRV impacted at approximately 100 feet per second and was seriously damaged.

Investigators determined that the main parachute disconnected just as it opened completely due to separation of all riser lines at their merge point, which was the link assembly. Normally, the link assembly disengaged from the parachute release mechanism when the vehicle load was transferred from the main chute/link assembly to the load line/release mechanism by shearing a locking rivet and allowing free separation of the link at a rotation angle of approximately 90 degrees. The rivet had not sheared, however, and the link retention mechanism was damaged.[20]

The recovery system installed on the F-15 RPRV employed a harness-transfer feature not found on other MARS equipment. Initially, as in standard MARS recoveries, the vehicle hung nose down while descending on the main parachute. However, in the NASA system, after a predetermined time, a transfer sequence moved the vehicle to a horizontal attitude. On the first 11 flights,

---

18. Ibid.

19. Ibid.

20. James R. Phelps, Harry R. Chiles, and William R. Petersen, "Report on F-15 RPRV Recovery Accident—Oct. 16, 1974," Dec. 18, 1974, NASA Flight Research Center, Edwards, CA, file location L1-7-5B-2.

harness transfer was programmed to occur 35 seconds after initiation of the recovery sequence. On subsequent flights, harness transfer was programmed to occur at 23 seconds in order to allow for more flight time and recovery at lower altitudes.

Analytical studies correlating flight load data with recovery system sequencing showed that the combination of approximately simultaneous harness transfer and main-parachute deployment could produce a sufficiently large force and rotation on the link assembly and release mechanism to separate the chute. Investigators determined that a rotation of about 100 degrees and a force of 45 pounds would easily pull the link assembly from the release mechanism without shearing the rivet. Harness release timing was deemed a contributing factor.[21]

As a result, investigators recommended reinstating the 35-second harness release timing in order to decrease g-loading. They also recommended that engineers devise a method for preventing rotation between the link assembly and release mechanism. An independent investigation by Teledyne Ryan Aeronautical, designer of the parachute disconnect mechanism, concurred with the board's findings.[22]

A Good Catch Saves the Day

Rather than repair the damaged vehicle, it was replaced with the second F-15 RPRV. During this craft's second flight, on January 16, 1975, research pilot Tom McMurtry successfully completed a series of planned maneuvers and then deployed the recovery parachute. During MARS retrieval, with the RPRV about 3,000 feet above the ground, the towline separated and the RPRV was once again in free flight but without anyone at the controls. McMurtry noticed that the RPRV had broken loose and become inverted. He managed to regain control of the vehicle while it was still 3,000 feet above the ground. Rolling the airplane upright, McMurtry sought to optimize his airspeed while guiding the RPRV toward an emergency landing on the Edwards PIRA. With little time and few choices, he made a straight-in approach. The vehicle slid in across the desert scrub, striking a Joshua tree and a raised ridge at the edge of a dirt road. Although the RPRV suffered some damage, McMurtry's actions saved it from destruction.[23]

---

21. Ibid.

22. Ibid.

23. Transcript of interview with Garrison P. Layton by Richard P. Hallion, n.d., circa 1977, DFRC Historical Reference Collection, NASA DFRC, Edwards, CA, file location L2-10-1A-5.

The F-15 RPRV was eventually modified with skids for landings on the dry lakebed at Edwards. NASA

As a result of this successful remotely piloted landing, and considering previous parachute recovery difficulties, further use of MARS was discontinued. The RPRV was modified with landing skids, and all flights thereafter ended with horizontal touchdowns on the lakebed.[24]

Spin Research and Contributions to Safety

The F-15 RPRV project came to a halt December 17, 1975, following the 26th flight, but this did not spell the end of the vehicle's career. Almost 2 years later, in November 1977, flights resumed under the Spin Research Vehicle (SRV) project. Researchers were interested in evaluating the effect of nose shape on the spin susceptibility of modern high-performance fighters. Flight testing with the F-15 model would augment previous wind tunnel experiments and

---

24. Project Document OPD 80-67, Spin Research Vehicle Nose Shape Project, Feb. 5, 1980, DFRC Historical Reference Collection, NASA DFRC, Edwards, CA, file location L3-8-4A-5.

Recovery personnel prepare to lift the F-15 SRV off the Edwards Air Force Base PIRA after the vehicle's spin chute became entangled with the pitot tube. NASA

analytical studies. Baseline work with the SRV consisted of an evaluation of the basic nose shape with and without two vortex strips installed. In November 1978, following nine baseline-data flights, the SRV was placed in inactive status pending the start of testing with various nose configurations for spin-mode determination, forebody pressure-distribution studies, and nose-mounted spin-recovery-parachute evaluation.

When flights resumed on February 18, 1981, the new nose parachute caught on the airplane's pitot tube after deployment. Einar Enevoldson guided the vehicle to a landing on the Edwards PIRA with only minor damage. No further problems were encountered during the remaining flights.[25]

When the SRV program ended in July 1981, the F-15 models had been carried aloft 72 times—41 times for the RPRV flights and 31 times for the SRV. A total of 52 research missions were flown with the two aircraft, 26 free flights with each. There had been only two ground aborts, one aborted planned-captive flight, and 15 air aborts (missions terminated prior to launch).

---

25. Ibid.

Of 16 MARS recoveries, 13 were successful. Five landings occurred on the PIRA and 34 on the lakebed.[26]

Flight data were correlated with wind tunnel and mathematical modeling results, and they were presented in the form of various technical papers. Tests of the subscale F-15 models clearly demonstrated the value of the RPRV concept for making bold, rapid advances in free-flight testing of experimental aircraft with minimal risk and maximum return on investment. R. Dale Reed wrote, "If information obtained from this program avoids the loss of just one full-scale F-15, then the program will have been a tremendous bargain."[27]

It was, indeed. Spin-test results of the F-15 model identified a potentially dangerous "yaw-trip" problem with full-scale F-15s equipped with an offset airspeed boom. Such a configuration, the F-15 RPRV showed, might exhibit abrupt departure characteristics in turning flight as angle of attack increased. Subsequently, during early testing of F-15C aircraft equipped with fuselage-hugging conformal fuel tanks (like those subsequently used on the F-15E Strike Eagle) and an offset nose-boom, Air Force test pilot John Hoffman experienced just such a departure. Review of the F-15 RPRV research results swiftly pinpointed the problem and alleviated fears that the F-15 suffered from some inherent and major flaw that would force a costly and extensive redesign. This lone "save" likely more than paid for the entire NASA F-15 RPRV effort.[28]

Lessons Learned

In what will become a recurring theme in the following mishap investigations, human factors contributed significantly to the outcome. Several points merit review:

- The primary cause of the accident was the crew chief's failure to properly assemble the release mechanism and the failure of the inspector to catch the error prior to flight. Improved training or better attention to detail might have prevented these mistakes.

---

26. Peter W. Merlin, "F-15 RPRV/SRV Flight Log," July 2001, DFRC Historical Reference Collection, NASA DFRC, Edwards, CA, file location L1-7-5B-2.

27. R. Dale Reed, "RPRVs—The First and Future Flights," *Astronautics and Aeronautics* (April 1974): pp. 26–42.

28. Recollection of Dr. Richard P. Hallion, who, as AFFTC Center Historian, was present at the postflight briefing to the Commander, Air Force Flight Test Center (AFFTC); see James O. Young, *History of the Air Force Flight Test Center, 1 January 1982–31 December 1982, v. 1* (Edwards AFB, CA: AFFTC History Office, 1984), pp. 348–352.

- Documentation of procedures for installing the device was inefficiently spread across several documents. Consolidation of relevant documentation would have prevented the need for crosschecking. Critical items should have been highlighted.
- Personnel experienced with the assembly and inspection procedures were unavailable at the time of the flight. Project managers should ensure that all primary and backup personnel working the flight are adequately trained and familiar with equipment and procedures.
- Although the Air Force MARS crew had experienced an identical incident, they had not shared this information with project personnel. It is important to communicate data that might be relevant to mission safety and success.
- A design flaw made it possible to install the release mechanism incorrectly. The manufacturer should have designed the device so that it could be installed only one way, but an astute observer (mechanic/inspector) also should have caught this hidden trap.

The HiMAT vehicle was a subscale representation of a notional future fighter design incorporating such features as canards, winglets, and supercritical wings. NASA

CHAPTER 2

HiMAT

Increasing Aircraft Performance Capabilities

In 1973, NASA and Air Force officials began exploring a project to develop technologies for advanced fighter aircraft. Several aerospace contractors submitted design proposals for a baseline advanced-fighter concept with performance goals of a 300-nautical-mile mission radius, sustained 8-g maneuvering capability at Mach 0.9, and a maximum speed of Mach 1.6 at 30,000 feet altitude. The Los Angeles Division of Rockwell International was selected to build a 44-percent-scale, remotely piloted model for a project known as Highly Maneuverable Aircraft Technology, or HiMAT. Flight-testing occurred at Dryden, initially under the leadership of project manager Paul C. Loschke and later under Henry Arnaiz.[1]

The HiMAT project represented a shift in focus by researchers at Dryden. Through the Vietnam era, the focal point of fighter research had been speed. In the 1970s, driven by a national energy crisis, new digital technology, and a changing combat environment, researchers sought to develop efficient research models for experiments into the extremes of fighter maneuverability. As a result, the quest for speed, long considered the key component of successful air combat, became secondary.

HiMAT program goals included a 100-percent increase in aerodynamic efficiency over 1973 technology, and maneuverability that would allow a sustained 8-g turn at Mach 0.9 and 25,000 feet altitude. Engineers designed the HiMAT aircraft's rear-mounted swept wings, digital flight-control system, and forward-mounted controllable canards to give the plane a turn radius twice as tight as that of conventional fighter planes. At near sonic speeds and an altitude

---

1. L.E. Brown Jr., M.H. Roe, and R.A. Quam, "HiMAT Systems Development Results and Projections," Society of Automotive Engineers Aerospace Congress and Exposition, Los Angeles, CA, Oct. 13–16, 1980.

of 25,000 feet, the HiMAT aircraft could perform an 8-g turn, nearly twice the capability of an F-16 under the same conditions.[2]

Subscale Test Bed

The scale factor for the RPRV was determined by cost considerations, payload requirements, test-data fidelity, close matching of thrust-to-weight ratio and wing loading between the model and the full-scale design, and availability of off-the-shelf hardware. The overall geometry of the design was faithfully scaled with the exception of fuselage diameter and inlet-capture area, which were necessarily overscale in order to accommodate a 5,000-pound-thrust General Electric J85-21 afterburning turbojet engine.

Advanced technology features included maximum use of lightweight, high-strength composite materials to minimize airframe weight; aeroelastic tailoring to provide aerodynamic benefits from the airplane's structural-flexibility characteristics; relaxed static stability to provide favorable drag effects due to trimming; digital fly-by-wire controls; a digital, integrated propulsion-control system; and such advanced aerodynamic features as close-coupled canards, winglets, variable-camber leading edges, and supercritical wings. Composite materials, mostly graphite/epoxy, comprised about 95 percent of exterior surfaces and approximately 29 percent of the total structural weight of the airplane. Researchers were interested in studying the interaction of the various new technologies.[3]

To keep development costs low and allow for maximum flexibility for proposed follow-on programs, the HiMAT vehicle was modular for easy reconfiguration of external geometry and propulsion systems. Follow-on research proposals involved forward-swept wings, a two-dimensional exhaust nozzle, alternate canard configurations, active flutter suppression, and various control-system modifications. These options, however, were never pursued.[4]

Rockwell built two HiMAT air vehicles, known as AV-1 and AV-2, at a cost of $17.3 million (by comparison, a single F-16 cost approximately $10.2 million in contemporary U.S. dollars). Each was 22.5 feet long, spanned 15.56 feet, and weighed 3,370 pounds. The vehicle was carried to a launch

---

2. Henry H. Arnaiz and Paul C. Loschke, "Current Overview of the Joint NASA/USAF HiMAT Program," NASA Conference Publication (CP) 2162, presented at the Tactical Aircraft Research and Technology Conference, Hampton, VA, Oct. 21–23, 1980.

3. Ibid.

4. Ibid.

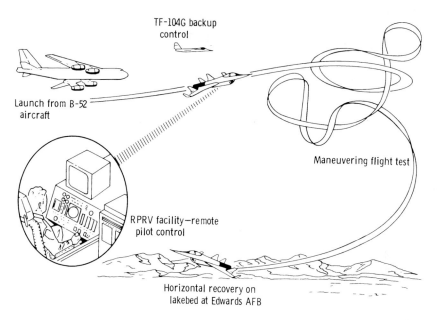

TF-104G backup
control

Launch from B-52
aircraft

Maneuvering flight test

RPRV facility—remote
pilot control

Horizontal recovery on
lakebed at Edwards AFB

The HiMAT was launched from a modified B-52. The primary pilot was located in a ground station, but a backup piloting capability was available in a chase plane. NASA

altitude of about 40,000 to 45,000 feet beneath the wing of NASA's B-52. Following release from the wing pylon at a speed of about Mach 0.7, the HiMAT dropped for three seconds in a preprogrammed maneuver before transitioning to control by the ground pilot. An airborne backup pilot— usually flight-test engineer Vic Horton—rode in the rear seat of a TF-104G chase plane and could take control of the HiMAT if necessary. Research flight-test maneuvers were restricted to within a 50-nautical-mile radius of Edwards and ended with landing on Rogers Dry Lake. The HiMAT was equipped with steel skid landing gear. Maximum flight duration varied from about 15 to 80 minutes, depending on thrust requirements, with an average planned flight duration of about 30 minutes.

As delivered, the vehicles were equipped with a 227-channel data-collection and recording system. Each RPRV was instrumented with 128 surface-pressure orifices with 85 transducers, 48 structural-load and hinge-moment strain gauges, six buffet accelerometers, seven propulsion-system parameters, 10 control-surface-position indicators, and 15 airplane motion and air data parameters. NASA technicians later added more transducers for a surface-pressure survey.

To guard against failure, redundant systems were incorporated throughout the vehicle. These included computers, hydraulic and electrical systems,

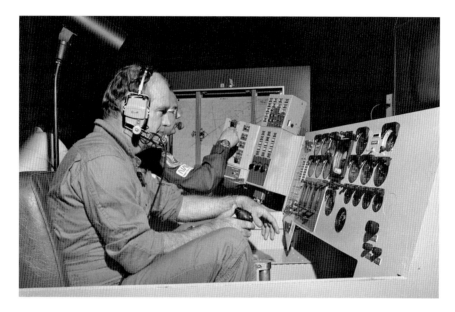

Bill Dana, foreground, controls the HiMAT from the ground cockpit while a flight-test engineer monitors the aircraft's systems. NASA

servo-actuators, uplink receiver/decoders and antennas, and downlink antennas. Flight-critical sensors, rate gyros, and accelerometers were triple-redundant.[5]

Flying the HiMAT from the ground-based cockpit using the digital-fly-by-wire system required control techniques similar to those used in conventional aircraft, although design of the vehicle's control laws had proved extremely challenging. The HiMAT was equipped with an innovative flight-test-maneuver autopilot based on a design developed by Teledyne Ryan Aeronautical, which also developed the aircraft's backup flight-control system (with modifications made by Dryden engineers).

For the first time ever, the autopilot system provided precise, repeatable control of the vehicle during prescribed maneuvers so that large quantities of reliable test data could be recorded in a comparatively short segment of flight time. Dryden engineers and pilots tested the control laws for the system in simulations and in flight, making necessary adjustments based on experience. Once adjusted, the autopilot was a valuable tool for obtaining high-quality, precise data that would not have been obtainable using standard piloting methods.

5. L.E. Brown, M. Roe, and C.D. Wiler, "The HiMAT RPRV System," American Institute of Aeronautics and Astronautics (AIAA) Paper 78-1457, presented at the AIAA Aircraft Systems and Technology Conference, Los Angeles, CA, Aug. 21–23, 1978.

The autopilot enabled the pilot to control multiple parameters simultaneously and to do so within demanding, repeatable tolerances. As such, the flight-test-maneuver autopilot was broadly applicable to flight research, and it offered potential benefit to other types of flight programs as well.[6]

The HiMAT vehicle was equipped with a backup control system (BCS) for emergency operation in the event of failure of the primary control system (PCS). The BCS allowed recovery of the vehicle from unusual or extreme attitudes, controlled vehicle dynamics throughout the flight envelope, and gave the ground pilot the capability to land the vehicle safely with minimal control input. The BCS was a full-authority, three-axis, multimode, multirate digital controller. It was programmed with stability augmentation and mode command functions. The various modes—automatically initiated depending on circumstances—included recovery, orbit, climb/dive, turn, landing, and engine-out. In the landing mode, airspeed and descent rate were keyed to radar altitude. Since the radar altimeter range was from zero to 5,000 feet above the ground, the pilot could select the landing mode once the vehicle was flying below 5,000 feet. Once the landing mode had been engaged, the system was capable of modulating both airspeed and altitude during approach, so the pilot did not need to make additional control inputs.[7]

Transfer to the BCS could be made manually from the ground cockpit or from controls in the rear cockpit of the TF-104G chase aircraft, or it could be made automatically in the event of certain system failures or loss of uplink or downlink signal carrier. Once in BCS mode, the onboard computer controlled the vehicle with limited discrete control input from either the ground-based cockpit or a flight-test engineer on board the chase aircraft.[8]

Initial Flight Testing

The maiden flight of HiMAT AV-1 took place July 27, 1979, with Bill Dana at the controls. All objectives were met despite minor difficulties. Most notably, a design flaw in the uplink signal receiver/decoder caused the vehicle to automatically transfer control to the BCS on two occasions. Engineers determined

6. E.L. Duke, F.P. Jones, and R.B. Roncoli, "Development of a Flight Test Maneuver Autopilot for a Highly Maneuverable Aircraft," AIAA Paper 83-0061, presented at the AIAA 21st Aerospace Sciences Meeting, Reno, NV, Jan. 10–13, 1983.

7. Robert W. Kempel and Michael R. Earls, "Flight Control Systems Development and Flight Test Experience with the HiMAT Research Vehicles," NASA Technical Paper (TP) 2822 (June 1988).

8. Arnaiz and Loschke, "Current Overview of the Joint NASA/USAF HIMAT Program."

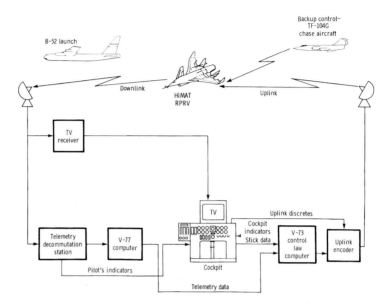

A second pilot in a two-seat TF-104G had backup controls for the HiMAT in case the PCS failed. NASA

that transfer to BCS resulted from either or both decoders receiving inadequate uplink signals as a function of vehicle attitude. To eliminate the problem, the hardware was redesigned using a diversity-combining concept that produced uninterrupted telemetry. The new system continuously combined the output signals of the dual receivers in proportion to their signal strength so that regardless of the airplane's orientation with respect to the transmitter, the best signal was available for all uplinked commands.[9]

The receiver was redesigned but the new equipment was not installed until the fourth flight. Several transfers to the BCS occurred on the second and third flights, but other than reducing the number of test points accomplished, they caused no real problems. Meanwhile, these incidents gave engineers the opportunity to verify proper functioning of backup controls.[10]

Despite these minor problems, the test flights resulted in the acquisition of significant data and cleared the HiMAT to a maximum speed of Mach 0.9 and an altitude of 40,000 feet, as well as demonstrating a 4-g turning capability. By the end of October 1980, the HiMAT team had flown the vehicle to Mach 0.925 and performed a sustained 7-g turn.

---

9. Kempel and Earls, "Flight Control Systems Development and Flight Test Experience with the HiMAT Research Vehicles."

10. Ibid.

By the fourth flight, the uplink signal receiver problem had been solved, and no transfer to BCS occurred. During one maneuver, however, fuel starvation resulting from a negative-g condition caused the engine to flame out. Ground pilot Bill Dana failed in his first restart attempt due to improper procedure, but he succeeded after recognizing the problem and trying again. He later commented—perhaps with some degree of sarcasm—that the ignition system should be "redesigned to allow air-starts to be a one-man procedure."[11]

PCS Failure

The fifth flight occurred on July 8, 1980, with Dana again serving as ground pilot. Just prior to takeoff, a computer providing guidance and ground-track data failed. Flight-test engineer Donald Gatlin reloaded the program, and the computer functioned normally for the remainder of the flight.[12]

The HiMAT vehicle was carried to an altitude of 45,000 feet beneath the wing of the B-52, accompanied by two TF-104G chase planes. After the HiMAT dropped away from the launch pylon, Dana performed steady-state sideslip maneuvers at Mach 0.8, Mach 0.85, and Mach 0.9 during descent to 25,000 feet. Using a chase plane as pacer, he found that his indicated airspeed (0.9 Mach) did not match that of the TF-104G (0.8 Mach). As Dana prepared to increase the HiMAT's speed to 0.95 Mach, the vehicle made an uncommanded transfer from its primary control system to its backup control system. Dana observed that several warning lights illuminated on his control panel, indicating failure of one of two uplink receiver electronic-signal-conditioning units (decoders).

The vehicle was now 50 nautical miles from Edwards, and mission controllers recommended against attempting to reengage the PCS. Dana executed a right turn to begin heading back to the base, reducing airspeed to Mach 0.76 in order to tighten the turn and remain within the planned work area. He also began dumping fuel, shutting off the flow with 340 pounds remaining in the tanks. After a chase plane crewed by Steve Ishmael and Vic Horton rendezvoused with the HiMAT, Dana started his descent toward an unpaved runway on Rogers Dry Lake.[13]

---

11. Paul C. Loschke, "HiMAT Flight Report, Flight H1-4-9, June 25, 1980," DFRC Historical Reference Collection, NASA DFRC, Edwards, CA, file location L1-8-3A-1.

12. Paul C. Loschke, "HiMAT Flight Report, Flight H1-5-10, July 8, 1980," DFRC Historical Reference Collection, NASA DFRC, Edwards, CA, file location L1-8-3A-2.

13. Ibid.

Observers aboard the TF-104G chase plane watch as the HiMAT slides across the dry lakebed on its belly. NASA

He was satisfied to find that BCS approach characteristics matched those he had experienced in the simulator. But on final approach, at an altitude of 1,000 feet, Dana encountered difficulty while attempting to extend the landing gear. Ishmael reported that the gear failed to deploy, so Dana executed a go-around and climbed to 5,000 feet. He cycled the gear switch during the 2-g turn, but still the gear failed to extend. Control of the HiMAT was transferred to Horton, who also attempted to lower the gear without success. Approximately 17 seconds later, with fuel running low, Dana took control of the HiMAT and performed a gear-up landing on the lakebed runway. The BCS provided semiautomatic landing capability with only slight heading input from the pilot. Touchdown occurred approximately 3,000 feet beyond the aim point at a speed of 185 knots, and the vehicle slid for 3,480 feet before coming to a stop.

The vehicle sustained substantial but repairable damage. Graphite skin on the left-hand aft corner of the nose-gear well and on the engine access door had delaminated. Both aluminum fairings aft of the nose-gear skid were destroyed. Several antennas on the belly, including those for the flight-termination system and uplink, were sheared off. The hydraulic bay air scoop was bent and part of the left vertical tail ventral fin ground off. Aside from damage incurred on touchdown, the only anomaly found during postflight inspection was a popped circuitbreaker for the receiver/decoder, the apparent failure of which led to the switch from primary to backup controls. Miraculously, considering the location of the air inlet on the underside of the vehicle, the engine had not ingested any debris.[14]

Due to the PCS failure and subsequent return to base, none of the primary mission objectives were accomplished, but the mishap revealed hardware and software discrepancies that might otherwise have gone undiscovered. Additionally, successful performance of the BCS validated the design and gave the HiMAT team confidence in their ability to fly the aircraft under backup controls on future flights, if necessary.

Accident investigators determined that metallic debris from a rubberized gasket with embedded metal slivers had caused the receiver/decoder failure that tripped the circuitbreaker, leading to failure of the PCS. The faulty unit, and others that incorporated the same type of gasket, were returned to the vendor to be cleaned, equipped with a different gasket (one with an embedded metal screen), and functionally checked.

Investigators recommended conducting failure-mode-effects testing of all uplink discrete commands. Tests would include scenarios in which one or both receiver/decoders had failed. Engineers discussed the possibility of furnishing the HiMAT control system with a downgraded primary mode of operation in the event of a receiver/decoder failure, but ultimately they decided that the backup control system was satisfactory.[15]

The circuitbreakers for the primary and backup receiver/decoders were upgraded from 2 and 3 amperes (amps) to 5 amps for each pair. For precautionary reasons, based on the criticality of the circuitbreakers to the control system, it was decided that maintenance technicians should periodically perform a pull test of all HiMAT circuitbreakers.

HiMAT landing gear deployment procedures and equipment were also revised after it was discovered that the landing gear deployment problem resulted from inadequate testing of newly released software. Engineers discussed providing the backup controller in the TF-104G with a rotary switch but opted instead to modify the procedure by having the controller leave the switch in each successive position for a full 2 seconds before making any changes. An emergency switch was added to the ground cockpit to bypass automatic encoding circuitry, allowing the pilot to deploy the landing gear in the event of failure of the primary system.[16]

---

14. Ibid.

15. Albert F. Myers, "Engineering Analysis Results of HiMAT Flight Mishap, July 8, 1980 (H1-5-10)," NASA DFRC E-EDC(AFM)3397 (Aug. 5, 1980).

16. Ibid.

Investigators examine the HiMAT following its belly landing. Fortunately, damage was not extensive. NASA

Contributions to Aircraft Technology

AV-2 was flown for the first time on July 24, 1981. The following week, Steve Ishmael joined the project as a ground pilot. After several airspeed calibration flights, researchers began collecting data with AV-2.

On February 3, 1982, Dana demonstrated 8-g maneuver capabilities with AV-1 for the first time. A little over 3 months later, researchers obtained the first supersonic data with the HiMAT, achieving speeds of Mach 1.2 and Mach 1.45. Research with both air vehicles continued through January 1983. Fourteen flights were completed with AV-1 and 12 with AV-2, for a total of 26 over 3½ years.[17]

The HiMAT research successfully demonstrated a synergistic approach to accelerating development of an advanced high-performance aircraft. Many high-risk technologies were incorporated into a single, low-cost vehicle and tested—at no risk to the pilot—to study interaction among systems, advanced materials, and control software. Design requirements dictated that no single failure should result in the loss of the vehicle. Consequently, redundant systems were incorporated throughout the aircraft, including computer microprocessors, hydraulic and electrical systems, servo-actuators, and data uplink/downlink equipment.[18]

---

17. HiMAT Flight Reports, 1979–1983, DFRC Historical Reference Collection, NASA DFRC, Edwards, CA.

18. Reed, "Flight Research Techniques Utilizing Remotely Piloted Research Vehicles."

The HiMAT program resulted in several important contributions to flight technology. Foremost among these was the use of new composite materials in structural design. HiMAT engineers used materials such as fiberglass and graphite-epoxy composites to strengthen the airframe and allow it to withstand high-g conditions during maneuverability tests. Knowledge gained in composite construction of the HiMAT vehicle strongly influenced other advanced research projects, and such materials are now used extensively on commercial and military aircraft. Designers of the X-29 employed many design concepts developed for HiMAT, including the successful use of a forward canard and the rear-mounted swept wing constructed from lightweight composite materials. Although the X-29's wings swept forward rather than aft, the principle was the same. HiMAT research also brought about far-reaching advances in digital flight-control systems, which can monitor and automatically correct potential flight hazards.[19]

Over the course of the test program, the two vehicles transferred automatically to BCS on 24 occasions. Vehicle AV-1 experienced 16 transfers, 12 during the first three flights. Only the transfer on the fifth flight was due to a hard system failure (the receiver/decoder), precluding transfer back to PCS and resulting in the only BCS landing during the entire program. Vehicle AV-2 experienced eight transfers to BCS, including one manual transfer. During cumulative BCS operation, researchers had the opportunity to exercise all BCS modes except engine-out, and all operated satisfactorily with no observed anomalies.[20]

Lessons Learned

The HiMAT PCS failure was caused by a mechanical malfunction. The landing gear malfunction stemmed from a software issue.

- The primary cause of the PCS failure was traced to a faulty receiver/decoder unit. Similar units were returned to the vendor for modification and testing, and investigators recommended failure-mode-effects testing of all uplink discrete commands.
- A procedural problem associated with modified flight software prevented the pilot from lowering the landing gear. This resulted from inadequate testing of newly released software.

---

19. HiMAT fact sheet (FS) 2002-06-025, NASA DFRC, Edwards, CA (June 2002).
20. Kempel and Earls, "Flight Control Systems Development and Flight Test Experience with the HiMAT Research Vehicles."

NASA researchers modified a Firebee II target drone, designated DAST-1, with supercritical wings of a shape optimized for a near-supersonic transport-type aircraft. NASA

CHAPTER 3

DAST

Exploring Aeroelastic Structural Design

In the early 1970s, researchers at Dryden and NASA Langley Research Center sought to expand the use of RPRVs into the transonic realm. The Drones for Aerodynamic and Structural Testing (DAST) program was conceived of as a means for conducting high-risk flight experiments using specially modified Teledyne-Ryan BQM-34E/F Firebee II supersonic target drones to test theoretical data under actual flight conditions. Described by NASA engineers as a "wind tunnel in the sky," the DAST program merged advances in electronic remote-control systems with advanced airplane-design techniques. The drones were relatively inexpensive and easy to modify for research purposes and, moreover, were readily available from an existing stock of U.S. Navy target drones.

The DAST researchers were most interested in correlating theoretical predictions and experimental flight results of aeroelastic effects in the transonic speed range. Such tests, particularly those involving wing flutter, would be extremely hazardous with a crewed aircraft.[1]

Correlating Theory with Flight Data

The unmodified Firebee II had a maximum speed of Mach 1.1 at sea level and almost Mach 1.8 at 45,000 feet, was capable of 5-g turns, and, in the basic configuration, provided baseline performance and handling data prior to installation of the research wing. Researchers modified two vehicles, DAST-1 and DAST-2, to test several wing configurations during maneuvers at transonic speeds in order to compare flight results with theoretical and wind tunnel findings. For captive and free flights, the drones were carried aloft beneath a Navy DC-130A or NASA's B-52. The DAST vehicles were equipped with

---

1. H.N. Murrow and C.V. Eckstrom, "Drones for Aerodynamic and Structural Testing (DAST)—A Status Report," presented at the AIAA Aircraft Systems and Technology Conference, Los Angeles, CA, Aug. 21–23, 1978.

Without an onboard TV camera, DAST pilots had to rely on the radar plot board to keep track of the vehicle's location. NASA

remotely augmented digital flight-control systems, research instrumentation, an auxiliary fuel tank for extended range, and MARS. On the ground, a pilot controlled the DAST vehicle from a remote cockpit while researchers examined flight data transmitted via pulse-mode telemetry. In the event of a ground computer failure, the DAST vehicle could also be flown using a backup control system in the rear cockpit of a Lockheed F-104B chase plane.[2]

The primary DAST flight-control system was remotely augmented. In this configuration, control laws for augmenting the airplane's flying characteristics were programmed into a general-purpose computer on the ground. Closed-loop operation was achieved through a telemetry uplink/downlink between the ground cockpit and the vehicle. Known as the remotely augmented vehicle, or RAV, concept, this technique had previously been tested using the F-15 RPRV.[3] Among the advantages of this technique were that the cost of a single control-

2. Ibid.

3. David L. Grose, "The Development of the DAST I Remotely Piloted Research Vehicle for Flight Testing an Active Flutter Suppression Control System," NASA Contractor Report (CR) 144881 (Feb. 1979).

system facility could be spread over multiple RPRV programs and control laws could be quickly changed without changes to flight hardware.[4]

Baseline testing was conducted between November 1975 and June 1977 using an unmodified BQM-34F drone. It was carried aloft three times for captive flights: twice by a DC-130A, and once by the B-52. These flights gave ground pilot Bill Dana a chance to check out the RPRV systems and practice prelaunch procedures. Finally, on July 28, 1977, the Firebee II was launched from the B-52 for the first time. Dana flew the vehicle using an unaugmented control mode called Babcock-direct. He found the Firebee less controllable in roll than simulations suggested it would be, but overall performance was higher. Dana successfully transferred control of the drone to Vic Horton in the rear seat of an F-104B chase plane, where he flew the Firebee through the autopilot to evaluate controllability before transferring control back to Dana just prior to recovery.

Technicians then installed instrumented standard wings in what was known as the Blue Streak configuration. Tom McMurtry flew a mission on March 9, 1979, to evaluate onboard systems such as the autopilot and RAV system. Results were generally good, with some minor issues to be addressed prior to flying the DAST-1 vehicle.[5]

The DAST-1 was fitted with a set of swept supercritical wings of a shape optimized for a transport-type aircraft capable of Mach 0.98 at 45,000 feet. The ARW-1 aeroelastic research wing, designed and built by Boeing in Wichita, KS, was equipped with an active flutter suppression system (FSS). Research goals included the validation of active controls technology for flutter suppression, enhancement and verification of transonic flutter prediction techniques, and creation of a database for aerodynamic-loads prediction techniques for elastic structures.[6]

Since it had no wing control surfaces, the basic Firebee drone was controlled through collective and differential horizontal stabilizer and rudder deflections. The DAST-1 retained this control system, leaving the ailerons free to perform the flutter suppression function that was not traditionally a feature on the Firebee. During fabrication of the wings, it became apparent that torsional stiffness was greater than predicted. In order to ensure that the flutter boundary

4. Dwain A. Deets and John W. Edwards, "A Remotely Augmented Vehicle Approach to Flight Testing RPV Control Systems," NASA Technical Memorandum (TM) X-56029 (Nov. 1974).

5. DAST flight logs and mission reports, 1975–1983, DFRC Historical Reference Collection, NASA DFRC, Edwards, CA, file location L1-6-4B-2.

6. John W. Edwards, "Flight Test Results of an Active Flutter Suppression System Installed on a Remotely Piloted Vehicle," NASA TM-83132 (May 1981).

This control room at Dryden provided additional capability for engineers monitoring DAST flights. NASA

remained at an acceptable Mach number, 2-pound ballast weights were added to each wingtip. These weights, known as tip masses, consisted of containers of lead shot that could be jettisoned to aid recovery from inadvertent large-amplitude wing oscillations. Researchers planned to intentionally fly the DAST-1 beyond its flutter boundary in order to demonstrate the effectiveness of the FSS.[7]

Along with the remote cockpit, there were two other ground-based facilities for monitoring and controlling the progress of DAST flight tests. Dryden's control room contained radar plot boards for monitoring the flightpath, strip charts indicating vehicle rigid-body stability and control and operational functions, and communications equipment for coordinating test activities. A research pilot stationed in the control room served as flight director. Engineers monitoring the flutter tests were located in the Center's Structural Analysis Facility (SAF). The SAF accommodated six people, with one serving as test director to oversee monitoring of the experiments and communicate directly with the ground pilot.[8]

---

7. Ibid.

8. Ibid.

Explosive Flutter

The DAST-1 was launched for the first time on October 2, 1979. Following the drone's release from the B-52, Tom McMurtry guided it through FSS checkout maneuvers and a subcritical-flutter investigation. An uplink receiver failure resulted in an unplanned MARS recovery about 8 minutes after launch. The second flight was delayed until March 1980. Again, only subcritical-flutter data was obtained, this time because of an unexplained oscillation in the left FSS aileron.[9]

The DAST-1 vehicle was lost in a mishap during the third flight, on June 12, 1980. Unknown to test engineers, the FSS was operating at one-half nominal gain, resulting in misleading instrument indications that concealed a trend toward violent flutter conditions at speeds beyond Mach 0.8.

Mission objectives included checkout of FSS operation in the subcritical region of the flight envelope and verification of control-law revisions. The aircraft was equipped with strain gauges and pressure transducers to record wing-loads data. The flight was to conclude with an operational checkout of the tip-mass system as well as acquisition of basic longitudinal-stability data at elevated g-levels. The desired test conditions were to be flown at 15,000 and 20,000 feet over a range of speeds from Mach 0.675 to Mach 0.90.[10]

Following the DAST-1's launch from the B-52, ground pilot McMurtry guided it to 15,000 feet and set up for the first test point, at Mach 0.675. Over the course of the flight, he successfully accomplished test maneuvers through Mach 0.775. In the Structural Analysis Facility at Dryden, John W. Edwards gave the ground pilot clearance to increase speed to Mach 0.825, the actual Mach number as opposed to the desired indicated Mach number of 0.80. As McMurtry increased his throttle setting, flight-test engineers Bill Andrews and Kurt Schroeder both commented that the flight plan called for Mach 0.80.

By this time, the DAST-1 was accelerating noticeably. When McMurtry requested verification of the intended Mach number, Edwards replied, "Slow down, slow down, slow down."[11] McMurtry decreased the throttle setting and applied back-pressure to the control stick, watching his instruments to ensure that engine revolutions per minute (rpm) were decreasing.

---

9. Peter W. Merlin, DAST flight logs, Sept. 1999, DFRC Historical Reference Collection, NASA DFRC, Edwards, CA, file location L1-6-4B-2.

10. Kenneth E. Hodge et al., "Investigation Report of DAST Mission Failure," Oct. 1980, DFRC Historical Reference Collection, NASA DFRC, Edwards, CA, file location L1-6-6A-1.

11. Ibid.

The DAST-1 experienced explosive flutter at Mach 0.825, leading to catastrophic failure of the right wing. NASA

With only one remaining wing, the DAST-1 plummeted toward the ground. NASA

By this time, video transmissions from a chase plane showed that the vehicle's right wing was fluttering wildly, and strip charts in the control room indicated low damper response to control input. As rapidly divergent oscillations saturated the FSS ailerons, Edwards called, "Terminate, terminate,"[12] an instruction to cease test maneuvers and decelerate as rapidly as possible. McMurtry jettisoned the wingtip masses, but this failed to arrest the flutter. Less than 6 seconds after oscillations began, the right wing broke into several pieces. In the ground cockpit, instruments indicated that the vehicle was rolling rapidly to the right. After seeing no response to attempts at regaining control, McMurtry activated the vehicle's parachute recovery system.

By the time McMurtry sent the normal recovery command, 15 seconds had elapsed since the wing failure. The DAST-1 vehicle had descended to 12,500 feet and was by then at Mach 0.77 and a dynamic pressure of 530 pounds per square foot. The drag chute had been designed for a normal deployment dynamic pressure of just 110 pounds per square foot. Additionally, with one wing missing, the vehicle's control system was unable to perform a pitch-up maneuver for deceleration. These deployment conditions resulted in structural failure of the main parachute's forward attachment, preventing full inflation of the main parachute canopy. The DAST-1 plunged toward the ground, shedding parts and finally crashing in low hills west of Cuddeback Dry Lake.[13]

Searching for Answers

Dryden Flight Research Center Director Isaac T. "Ike" Gillam IV convened an investigation board the following day to determine the cause of what was termed a mission failure rather than an accident.[14] Members included Chairman and Director of Engineering Kenneth Hodge, along with engineers Gary Layton, Eldon Kordes, Moses Long, and Perry Hanson, as well as James Neher from Dryden's Office of Safety and Quality Assurance. The board com-

---

12. Ibid.

13. H.T. Jones, memorandum to D.J. Mourey, "DAST 1 Accident Investigation Report," Teledyne Ryan Aeronautical, July 10, 1980, DFRC Historical Reference Collection, NASA DFRC, Edwards, CA, file location L1-6-6A-1.

14. NASA Procedural Requirements document NPR 8621.1B defines a mission failure as a mishap of whatever intrinsic severity that prevents the achievement of a mission or project's minimum success criteria or minimum mission objectives as described in the mission operations report or equivalent document.

Because high-speed deployment conditions prevented full inflation of the main-parachute canopy, the DAST-1 was badly damaged when it struck the ground. NASA

pleted its investigation within 3 months, but Neher chose not to sign it because he felt that the final report failed to fulfill the board's charter.[15]

Key areas of investigation included an examination of Dryden flight-test procedures, the observed Mach number discrepancy, the active flutter suppression system, tip-mass-system performance, recovery system performance, and the failed wing structure. Results were based on telemetry records, taped recordings of voice communications, and witness statements.

Dryden flight-test procedures for the DAST began with the selection of the desired test conditions for each data point and determination of the number of data points required. Flight planners used a real-time simulation to establish the sequence of flight events, proposed flight track, and time required for each test point. During the actual flight, the DAST pilot maneuvered the

---

15. Kenneth E. Hodge, memorandum to Ike Gillam, "Subject: Investigation Report of DAST Mission Failure," Jan. 15, 1981, DFRC Historical Reference Collection, NASA DFRC, Edwards, CA, file location L1-6-6B-6.

vehicle to the planned airspeed and altitude conditions. Once test conditions had been stabilized, an experimenter in the control room initiated symmetric and asymmetric frequency sweeps using the flutter suppression system. While telemetered data were being analyzed with regard to response and damping, the pilot executed a series of control pulses. Response to these pulses was monitored on strip charts.[16]

The analyzed results were then compared with predicted data for the flight conditions from previous analyses and tests. If results were as expected and the damping response satisfactory, the experimenter cleared the vehicle for the next test point. While accelerating toward the next set of test conditions the pilot again executed a series of control pulses for analysis in order to estimate any changes that might occur due to acceleration to a higher Mach number. Upon reaching the desired conditions at the next test point, the process was repeated. This procedure incorporated careful planning of each test point, and it allowed researchers to evaluate a large number of test points during each flight.

Investigators concluded that such an incremental procedure, applying both wind tunnel and flight-test data, was consistent with best practices for this type of test operation. All critical parameters for evaluation of the vehicle's flutter response were available for real-time monitoring and near-real-time analysis. The large number of planned test points, however, imposed a heavy workload on personnel monitoring the flight. This resulted in delayed computations of frequency and damping, with results not being immediately displayed to all control room team members. Key personnel, such as the test engineer, had all the necessary information before giving clearance to proceed to the next test point, but Mach number and dynamic pressure values were not prominently displayed in the control room so that all personnel responsible for monitoring and analyzing the data could readily observe these parameters.[17]

During the flight, only one of the chase pilots and a few observers on the ground were aware of a Mach number discrepancy. The DAST pilot's instruments showed an airspeed value that was Mach 0.02 lower than the indications in the T-38 television chase and in the Spectral Analysis Facility at Dryden. As a result of this discrepancy, the pilot missed a planned low-speed test point. Investigators concluded that this was due to a calibration error. The Mach number discrepancy was not considered a major factor in the mishap.

Results from the first flight of the DAST-1 vehicle revealed that instability modes played a significant role in the performance of the FSS. Because data from this flight indicated that the original design did not provide sufficient

---

16. Hodge et al., "Investigation Report of DAST Mission Failure."
17. Ibid.

flutter control, engineers made improvements to the system. Software programmers changed the FSS control laws to provide improved flutter protection. When the improved FSS was first used during Flight 3, researchers found that it was unable to counteract the explosive flutter that occurred under conditions in which it had not been expected.[18]

The DAST-1 vehicle experienced a rapidly divergent symmetric wing-flutter mode at a Mach number and altitude that were within a region of the flight envelope predicted to be flutter-free with the FSS operating at design gains. At the time of the mishap, however, FSS gains were at only 50 percent of the design values established by a control law analyst prior to the flight. Postflight examination of FSS block diagrams provided by Boeing to NASA revealed that a required accelerometer input signal attenuation factor of 0.5 (for symmetric flutter) had been incorporated twice, effectively reducing system gains by half. Under these circumstances, the level of flutter damping afforded by the FSS at the flight conditions where the incident occurred was less than predicted for nominal gains. Investigators determined that the error occurred inadvertently when Boeing converted the analytical design to a block diagram notation. This error, the primary cause of the mishap, resulted in a configuration that caused the supercritical wing to become unstable at lower Mach numbers than anticipated, causing the vehicle to experience closed-loop flutter. Although NASA specifications required the FSS to perform satisfactorily between 0.5 and 2.0 times the nominal gain level, Boeing and NASA analyses conducted using updated mathematical models predicted that under some flight conditions, the specified criteria would not be met.[19]

The explosive nature of the flutter mode rendered the tip-mass release system ineffectual. The tip-mass system was designed so that releasing the weight from each wingtip moved the flutter boundary one-tenth of a Mach number higher than would be the case with the mass attached. Engineers designed the system to automatically release the mass when flutter amplitude reached preset conditions or to release it manually from the control room. Both methods were used during the DAST-1 mishap sequence, but the FSS failed to alleviate the flutter mode. Wind tunnel tests at the Langley Research Center had demonstrated that releasing the two weights from the wingtips could arrest flutter modes experienced by conventional airfoils. The DAST vehicle, equipped with the ARW-1 supercritical airfoil, suffered violent flutter divergence that was not arrested by tip-mass release or by subsequent separation of the wingtip and tip-mass dispensing system. The divergence encountered during Flight 3 was

---

18. Ibid.
19. Ibid.

The DAST-1 vehicle is seen here following a typical MARS recovery. NASA

an order of magnitude more than had been encountered in wind tunnel tests. Investigators determined that although the tip-mass system apparently worked as designed, the effectiveness of the system was not as predictable and did not stop the wing flutter even after the wingtip separated completely.[20]

Finally, the vehicle's parachute recovery system failed due to deployment outside its design envelope. Although the mechanism had demonstrated high reliability under nominal conditions, structural damage incurred during the mishap sequence led to premature deployment of the main parachute at higher than normal speeds. The ensuing rapid rate of descent precluded successful MARS recovery, and the vehicle suffered extensive damage at impact. The DAST mishap investigation board concluded that the recovery system and procedures were not adequate for recovery of damaged vehicles.

Examination of the failed wing structure provided additional information. Video footage, strain gauge data, and detailed studies of recovered components revealed that structural failure of the wing was progressive and followed the onset of flutter. The flutter control surfaces departed first, followed by both wingtip

---

20. Ibid.

The DAST pilot's control panel included all the basic cockpit instruments. A television screen provided the only outside visual reference. NASA

assemblies. The lower surface of the right wing tore away with the wingtip, resulting in reduced rigidity of the airfoil. Subsequent increasing oscillations broke the bolts that attached the wing spar to the fuselage. Metallurgical examination of these bolts indicated ductile failure with no evidence of prior fatigue.[21]

Although classified as a mission failure, Flight 3 nonetheless provided valuable information and lessons. Collected data helped establish state-of-the-art techniques for flutter prediction in the transonic speed regime and improved the accuracy of these techniques for flutter boundary prediction at a given altitude. Engineers obtained in-flight performance data for the flutter suppression system for application to future development of predictive methods and flutter control mechanisms. Project managers established improved configuration control procedures for research systems operated by geographically separated

---

21. Ibid.

teams. Additionally, engineers obtained high-quality data on the performance of an aeroelastically tailored wing for comparison with future development of predictive techniques.

The investigation board made several recommendations, beginning with the need to treat mission-critical items, such as the FSS, as complete end-to-end systems. Upon implementation of a new system or change to an existing one, investigators recommended that an analyst should conduct sensitivity studies and system impact assessments and specify test procedures and expected results from an end-to-end systems test. Any design work should be subject to strict configuration control procedures. Systems analysts, designers, and the flight-test team should review the design, execution, and results of the end-to-end systems test. Whenever possible, system performance should first be evaluated in flight at subcritical conditions before proceeding to supercritical conditions.[22]

Investigators also suggested that effective displays of such critical parameters as Mach number and dynamic pressure be provided to all persons involved in real-time test monitoring. Specific to the DAST program, they also recommended reevaluating flight-test procedures used during approach to critical aeroelastic conditions. Particular areas of concern included the size of the Mach/airspeed increment between successive test points, the portion of the flight envelope over which initial FSS evaluations were conducted, and the degree of reliance on the FSS for arresting flutter. There was some discussion regarding the need to tailor recovery system sequencing and procedures to the DAST flight envelope with consideration for recovering a damaged vehicle. Board members cautioned that any proposed changes should not degrade system reliability or require a development program in which costs would be out of proportion to expected benefits.[23]

Dissenting Opinion

James Neher from the Dryden Office of Safety and Quality Assurance served as investigation board secretary. He refused to sign the final report, and he offered a few principal objections. In a September 16, 1980, memorandum to board president Kenneth Hodge, he wrote, "I believe the attention and resulting rhetoric over the half-gain problem is severely hindering the attention which

---

22. Ibid.

23. Ibid.

should be given to a much greater problem, the problem of recognizing during flight when catastrophe is imminent but preventable."[24]

Despite the best efforts of future researchers, he asserted, there would always be uncertainty in flight-testing. Neher suggested reducing this uncertainty through improved end-to-end tests and other means while acknowledging that some critical detail might yet be missed. "Flight monitoring," he wrote, "is the last resort for catching the error before it bites us."[25] He noted that often in the case of a mishap, investigators might argue at length over whether a pilot's decision was correct when in fact the pilot had had to make the debatable decision in a brief moment while subjected to extreme stress. Neher recommended focusing on ascertaining what displays and warning mechanisms would more nearly ensure that a split-second decision made under duress would be the correct one.

Applying these principles to the DAST-1 mishap, he made several recommendations. First, he advocated that data-processing equipment with faster response time as well as better presentation and warning information be provided for the ground pilot and control room personnel. He also suggested direct, continuous presentation of damping ratio, and audible and visual warning signals whenever immediate termination was warranted based on a specified damping ratio threshold. In addition to improving vehicle safety, he noted, such improvements would also allow researchers to expedite flutter envelope clearance and reduce the vehicle's exposure to flutter conditions.

Neher stressed the need for airspeed accuracy and altitude control, and he emphasized that calibration discrepancies should not be tolerated. He suggested, again, that data processing and presentation played a key role in these issues.

With regard to the explosive flutter mode, he could only offer that researchers should attempt to develop a calculated structural change or improved FSS to safely arrest explosive flutter. He recognized that pilots should be aware that airspeed reduction to arrest flutter is only effective under some conditions.[26]

Dryden Director Isaac Gillam implemented a number of changes in response to the investigation report and Neher's comments. These included acceptance of suggestions regarding improved test procedures and systems checkout, improved display and presentation of real-time data, reevaluation of procedures for critical aeroelastic testing, improved instrument calibration, and techniques for better control of the vehicle's speed and altitude. He did not

---

24. James A. Neher, memorandum to Kenneth E. Hodge, "Subject: DAST Mission Failure Report," Sept. 16, 1980, DFRC Historical Reference Collection, NASA DFRC, Edwards, CA, file location L1-6-6B-6.

25. Ibid.

26. Ibid.

The DAST-2 was equipped with the improved ARW-1R supercritical wing. NASA

approve significant changes to the recovery system, as he did not consider doing so a prudent expenditure of the Center's resources. Instead, he directed the DAST engineering team to incorporate a speed brake into the vehicle. The FSS tip-mass system was modified for quicker response, but the remaining potential for explosive flutter and catastrophic loss was deemed an acceptable risk.[27]

"Alfalfa Impact Study"

Subsequently, the experimental ARW-1 wing was rebuilt as the ARW-1R and installed in a second DAST vehicle in order to continue the research program. The DAST-2 underwent initial captive-systems checkout beneath the wing of the B-52 on October 29, 1982, followed by a subcritical-flutter envelope expansion flight 5 days later. Unfortunately, the flight had to be aborted early due to unexplained wing structural vibrations and control-system problems.

---

27. Isaac T. Gillam IV, memorandum to NASA Headquarters, "Subject: Report of the DAST Mission Failure Investigation Board," July 22, 1981, DFRC Historical Reference Collection, NASA DFRC, Edwards, CA, file location L1-6-6B-6.

In 1983, a DC-130A was used as a mother ship for the DAST-2. Following two captive flights, the first launch attempt ended in failure. NASA

The next three flight attempts were also aborted, the first due to a drone engine temperature warning, the second because of the loss of telemetry, and a third for unspecified reasons prior to taxi.[28]

Further testing of the DAST-2 vehicle was conducted using a Navy DC-130 launch aircraft operated by Lockheed Aeronautical Services Company. Following two planned captive flights for systems checkout, the vehicle was declared ready to fly the 18th DAST research flight.[29]

On June 1, 1983, the DC-130A departed Edwards in a climbing turn over Mojave and California City, passing Cuddeback Lake and Barstow, CA, before turning west. Rogers Smith flew a TF-104G with backup pilot/flight-test engineer Ray Young while Einar Enevoldson began preflight preparations

28. Edwards, "Flight Test Results of an Active Flutter Suppression System Installed on a Remotely Piloted Vehicle."

29. Merlin, DAST flight logs.

from the ground cockpit. On previous flights, all uplinked commands to the DAST vehicle had been locked out for a period of 3 seconds to allow for sufficient vertical separation between the drone and the carrier aircraft before the ground pilot resumed control. This time, however, prior to launch, the ground cockpit was configured with all control surfaces in active mode.[30]

For Flight 18, uplinked commands to the horizontal tail surfaces were set at 2.0 degrees trailing-edge down, and the rudder was set at zero degrees deflection. Launch occurred over Harper Dry Lake, about 35 miles from Edwards, at an altitude of 18,000 feet. Immediately after separation from the launch pylon, the drone's recovery system drag chute deployed, but the main parachute was jettisoned while still packed in its canister.[31]

The drone plummeted to the ground in the middle of a farm field west of the lakebed. It was completely destroyed but, other than the loss of a small patch of alfalfa at the impact site, there was no property damage. Much later, when it was possible to joke about such things, a few wags referred to this event as the "alfalfa impact study."[32]

An investigation board found that a combination of several anomalies—a design flaw, a procedural error, and a hardware failure—simultaneously contributed to loss of the vehicle. These included an uncommanded recovery signal produced by an electrical spike, failure to reset a drag-chute timer, and improper grounding of an electrical relay. Another section of the investigation focused on project management issues. Criticism of Dryden's DAST program management was hotly debated, and several dissenting opinions were filed along with the main report.[33]

Since the DAST vehicle had separated from the pylon in recovery mode, the onboard autopilot was already engaged and initiated the recovery sequence. The engine fuel valve was simultaneously shut off and, since a 10-second timer in the recovery logic loop was timed out, the drag chute deployed immediately after launch. The main chute normally deployed at an altitude of 15,000 feet for a MARS recovery, but in this instance it was found still packed in its canister.

---

30. Glenn Sakamoto, memorandum to DAST Project Manager, "Subject: DAST Flight 18 Autopilot Evaluation," June 21, 1983, DFRC Historical Reference Collection, NASA DFRC, Edwards, CA, file location L1-6-4B-3.

31. Paul C. Loschke, Garrison P. Layton, George H. Kidwell, William P. Albrecht, William H. Dana, and Eugene L. Kelsey, "Report of DAST-1R Test Failure Investigation Board," Dec. 30, 1983, DFRC Historical Reference Collection, NASA DFRC, Edwards, CA, file location L1-5-4-15.

32. This event is recorded with other DAST mission markings painted on the side of the NB-52B even though the modified Stratofortress was not the launch aircraft for the ill-fated mission.

33. Loschke et al., "Report of DAST-1R Test Failure Investigation Board."

Investigators examine the DAST-2 wreckage at the crash site. The primary cause was traced to an electrical problem. NASA

The explosive bolts in the riser-release system, which normally would have fired after parachute deployment, had detonated inside the canister.

The DAST Test Failure Investigation Board determined that a variety of problems needed resolution. Otherwise, the project would likely continue to experience inefficiency, low flight rate, and a high probability of subsequent test failures.[34]

Loss of the vehicle was attributed to the inadvertent firing of the main parachute riser-release pyrotechnics. This anomaly occurred simultaneously with the deployment command, disconnecting the main chute from the vehicle and allowing it to crash. Two other technical anomalies preceded this event. First, unknown to the launch crew, the vehicle was already in recovery mode. Second, the delay timer failed to implement the normal 10-second delay. The most probable cause of the release mechanism failure was a latent short to ground in the wiring between riser-release relays in the power distribution box and the impact switch in the keel. Investigators determined that, due to

---

34. Ibid.

the nature of this fault, the vehicle would have been lost even at the end of a successful mission upon issuance of a normal recovery command. The recovery command at launch was likely due to a failure in the circuit used to sense generator failure and provide that information to the logic card. No specific cause was determined for failure of the 10-second timer. Investigators recommended performing a prelaunch test of the recovery system below 15,000 feet to verify that the riser-release circuitry was functioning properly.

Electrical spikes introduced into the logic card during umbilical connector separation at launch were seen as a probable cause of the uncommanded recovery signal. During previous flights, all control and recovery commands were inhibited for 3 seconds after launch. Since, for Flight 18, the logic card had been modified to permit active controls at launch, it seemed likely that such spikes might have occurred previously but had not been detected due to the positive-inhibit circuits. The board recommended performing simulated launch tests, including separation of the DAST umbilical from the launch aircraft, to assure electromagnetic compatibility of the logic card with the vehicle's electrical system and controls.[35]

It seemed clear to investigators that the DAST vehicle suffered from one or more insidious electrical or electronic problems that were poorly understood. Investigators recommended conducting a thorough study and test program to identify any sneak circuits or ground loops prior to flight.

Prior to Flight 18, all DAST launches had been accomplished using the B-52. Use of the DC-130 as a launch aircraft introduced unknowns into the prelaunch tests and created uncertainties. Lockheed Aeronautical Services Company operated the aircraft from Ontario Airport in California. Because the DC-130 was normally used for non-NASA missions and maintained by non-NASA personnel at a remote location, investigators noted that it was difficult to ensure its integrity prior to the DAST flight, largely due to time and cost constraints.[36]

Dryden quality assurance specialist Glenn Angle inspected the DC-130 launch system. He noted that, "Although Lockheed's procedures for documenting events and their inspection verification procedures are not what we at Dryden would consider to be good systems, they do appear to be adequate for what they do and their record of target drops on time speak well of Lockheed."[37]

---

35. Ibid.

36. Ibid.

37. Glenn E. Angle, memorandum to Jim Cooper, OP/C-130 Team Leader, "Subject: DAST Test Failure Investigation," June 30, 1983, DFRC Historical Reference Collection, NASA DFRC, Edwards, CA, file location L1-6-4B-4.

Launch of the DAST-2 was controlled from this station aboard the DC-130A. NASA

The board recommended that all future DAST launches be conducted from the NASA B-52. This would insure adherence to NASA maintenance practices and policies.[38]

The DAST failure investigation also revealed that generation and approval of test/operations documentation lacked consistency and rigor. Few of the numerous DAST test procedures had received formal managerial approval. Flight checklists had no standardized format, and a critical Recovery Reset step had been omitted. Some checklist items were found to be confusing and inconsistent. Investigators also found no clear, concise design documentation for DAST systems. As a result of these findings, they recommended instituting a disciplined process for documenting test operations that included a formal system for developing, reviewing, and approving procedures and checklists.

For the sake of simplicity and affordability, the DAST vehicle lacked redundant systems. Analysis showed that, from launch through the flight envelope and recovery, a single failure could result in loss of the vehicle. The recovery system and associated electronics and power systems were exceedingly complex by the standards of the time, and they included components manufactured as much as two decades earlier. The board asserted that with "the complexity of the system, the antiquity of the electronics, and the single failure design philosophy, the probability of completing a successful mission is low."[39]

Board members recognized that the DAST research wing assemblies and systems were designed to be single-string and could not be made redundant. They recommended modifying the DAST recovery system to reduce complexity and assure that no single circuitry failure would cause loss of the vehicle. Additionally, they suggested publishing a Statement of Accepted Risk addressing the necessity of accepting a certain amount of risk when flight testing a high-cost research experiment in which the test vehicle may be subject to loss from a single-point failure.

---

38. Loschke et al., "Report of DAST-1R Test Failure Investigation Board."
39. Ibid.

The DAST test team may have missed the potential for electrical/electronic problems simply because there was no single set of electrical system drawings that accurately represented the configuration used in the vehicle. A collection of DAST electrical system diagrams included a loose assembly of copies from technical manuals, engineering sketches attached to work orders, and accumulated composite drawings incorporating miscellaneous changes. This disorganized system resulted in a high potential for error. The board recommended development of a complete set of accurate drawings to be maintained under a disciplined configuration control system.

More Dissenting Opinions

In what became the most controversial element of the board's final report, investigators charged that management practices contributed significantly to problems with the DAST. Interviews with key personnel outlined the perceptions of several individuals with regard to project coordination, support, and scheduling problems. Concluding remarks in the report identified several factors "contributing to the poor flight record and low morale of the project team."[40]

Those interviewed cited high turnover in the position of project manager, untimely reassignments of several key avionics and control-system engineers, project personnel assigned to multiple high-priority projects, managers' inconsistent interpretations of project priorities, lack of commitment to schedules, and lack of a process for providing accurate and timely cost-time data. Investigators recommended organizing an experienced and adequately staffed project team and keeping it intact, formally defining the priority of the DAST program, and devising a method for retrieving current and historical cost-time information for use in planning, scheduling, and tracking all levels of management.[41]

Several individuals submitted comments on the board's findings and recommendations. DAST project manager Henry Arnaiz had no objections to the technical findings but felt that a recommendation to remove all capability to fire the parachute riser-release pyrotechnics after launch might significantly endanger life and property. He noted that a missed MARS recovery could result in the DAST vehicle being dragged by the wind following touchdown. He also felt that it was unnecessary to prohibit use of the DC-130 as an alternate launch aircraft since there was no evidence that any element of the DC-130 operation was responsible for the mishap. Arnaiz also stated that there was no

---

40. Ibid.
41. Ibid.

evidence that inadequate management practices had contributed to the riser-release-system malfunction.[42]

DAST principle investigator Glenn Gilyard took exception to a number of the board's comments and the overall tone of the report. Outlining his concerns in a memorandum to Dryden managers, he was particularly distressed by repeated references to "low morale of the DAST project team."[43] He felt that these comments were misleading, inflammatory, and unsubstantiated. He noted that comments of this type came from "statements made by a few, seemingly disgruntled individuals,"[44] and that the program manager, principle investigator, and several other key DAST personnel were never interviewed. Gilyard saw little, if any, direct or indirect connection between the management findings and the technical findings cited by the DAST-2 Test Failure Investigation Board. He charged that "conjecture, unsubstantiated statements, innuendo, and generalizations, that are interspersed throughout the report, detract from the worthy technical findings and recommendations."[45] He also expressed his opinion that despite the DAST program's problematic history, "the project team performed professionally."[46]

As with the earlier DAST-1 mishap investigation, James Neher offered the most scathing comments. He noted that Dryden's method of operation had resulted in many notable achievements in flight research but charged that projects with sporadic activity, few flights, and little management support—particularly for the DAST program—had "embarrassed and tarnished the center's reputation."[47] He wrote that change was warranted but warned, "It would be an injustice to let the ills of Firebee/DAST cause such an overreaction that advantages of the Dryden operation are negated."[48]

In an effort to develop lessons from what he referred to as "the inglorious record of Firebee/DAST," Neher submitted his own set of findings and

---

42. Henry Arnaiz, memorandum to Chief Engineer, "Subject: Review of the DAST-1R Test Failure Investigation Board Report," Feb. 1, 1984, DFRC Historical Reference Collection, NASA DFRC, Edwards, CA, file location L1-6-4B-4.

43. Glenn B. Gilyard, memorandum to OF/Chief, Flight Support Division, "Subject: Review of Report of DAST-1R Test Failure Investigation Board," Feb. 1, 1984, DFRC Historical Reference Collection, NASA DFRC, Edwards, CA, file location L1-6-4B-4.

44. Ibid.

45. Ibid.

46. Ibid.

47. James Neher, "Analysis—Firebee Accident Board," Sept. 1, 1983, DFRC Historical Reference Collection, NASA DFRC, Edwards, CA, file location L1-6-4B-4.

48. Ibid.

recommendations in which he suggested that such interrelated issues as planning, priorities, scheduling, communication, staffing, skills mix, teamwork, training, control room discipline, morale, and project philosophy needed to be addressed.

His first finding examined the philosophy behind the use of remotely piloted aircraft as research vehicles. Initially, the Firebee drone had been selected for the DAST program in the belief that it offered a quick and reasonably inexpensive alternative for a task deemed too hazardous for a crewed research aircraft. In practice, however, even off-the-shelf vehicles such as the BQM-34 proved more difficult to operate than anticipated. The vehicle's small size also imposed a limitation on the amount and scale of instrumentation that could be installed to collect data.[49]

Neher suggested that experiments such as the aeroelastic research wing could be piggybacked onto larger crewed aircraft, perhaps extending the experimental device from the bomb bay of a B-52 or attaching it to the wing or fuselage of a suitable platform. He also offered the possibility of using an optionally piloted aircraft in which a pilot would be on board for all checkout and nonhazardous experimentation. At the time, QF-100 drones—1950s jet fighters converted to optionally piloted aircraft—were available for $235,000 each, a relative bargain. According to Neher, use of the B-52, QF-100, or other similar aircraft would eliminate the need for a recovery aircraft (as used with MARS), thus reducing mission cost and complexity. The larger aircraft would also resolve issues such as lack of equipment space and the difficulty inherent in visual tracking of the aircraft.

Neher claimed that "lack of an effective team approach to project operation"[50] was a root problem of Firebee/DAST. Too many people, he argued, were assigned to multiple projects or simply reassigned and replaced. "Furthermore, the project manager had too little control over assigned people and was unable to keep members working together, full time as a team, in a manner properly conducive to adequate safety."[51]

He recommended that Dryden managers establish rigid criteria for assignment of personnel. The Center director would authorize changes only after careful consideration of the impact that such changes would effect. Neher also advocated that project managers be given increased authority over project members, perhaps offering incentives for achievement.

He condemned what he saw as inadequate planning, prioritization, and scheduling discipline. He felt that higher-priority projects "robbed Firebee/

49. Ibid.
50. Ibid.
51. Ibid.

DAST of resources and otherwise disrupted work and schedules."[52] Neher urged Dryden managers to reduce overindulgence of high-priority projects while correspondingly stabilizing resources and schedules of lower-priority projects. He suggested reducing the number of projects at the Center, if necessary, to the degree that each could be professionally staffed and scheduled, "being neither a usurper of resources assigned to other projects nor a project constantly decimated."[53]

Neher feared that top technical experts were being promoted to managerial positions and replaced with less experienced personnel, affecting both schedule and safety. He recommended that upper management consider promoting individuals with an eye toward reducing losses of qualified technical people to administrative jobs.

He also charged that the Center's closeout processes for design/readiness reviews were inadequate. He suggested improvements to the process, including designation of appropriate project managers; preparing closeout documents; and having design/readiness review team members, safety representatives, the chief engineer, and other select senior managers submit comments to the Center director prior to signature approval.

Between 1974 and 1983, according to Neher, considerable effort had been expended to upgrade Dryden's drawing and configuration control directives, but that progress had been impeded by reorganization (or rumors of impending reorganization) and less stringent legacy practices. He also complained of checklist errors, inadequate control room training, and insufficient accounting of man-hours expended in the accomplishment of various tasks. Neher recommended reviewing and improving these processes and methods of expediting communication of critical information to the Flight Readiness Review team.[54]

Dryden managers also offered comments on the investigation board's report. Chief engineer Milton O. "Milt" Thompson drafted a lengthy rebuttal, complaining that "it appears that the board is usurping the program and project manager's prerogative in defining specific solutions or corrective actions for each finding."[55]

He also noted that managers strongly disagreed with the board's implication that little serious consideration had been given to the selection of the Firebee as a baseline vehicle for the DAST program. "The Firebee was selected…by

52. Ibid.
53. Ibid.
54. Ibid.
55. Milton O. Thompson, "DAST Rebuttal," draft copy and notes, 1983, Milton O. Thompson papers, DFRC Historical Reference Collection, NASA DFRC, Edwards, CA, file location L1-5-4-12.

both Langley and Dryden since it was available at no cost, it had the required performance, it was operational with both the Navy and USAF, and it was considered to be ideally suited for use as an expendable high-risk test bed."[56]

Thompson felt that the overall program philosophy of using a simple vehicle with minimal modifications was generally sound, except for the failure to consider the need for a backup research wing assembly. He pointed out that five Firebee drones had been obtained "because we collectively expected to lose a vehicle,"[57] but only one experimental wing had been constructed.

According to Thompson's draft, the strongest disagreement lay with the board's blanket criticism of the Center's management team. The first major point of contention was investigators' complaints that DAST project managers suffered from continuous turnover of key project personnel; a new project manager had been assigned six times since 1973. "What the board failed to mention is that during that same time period the center had four different center directors or site managers, seven different deputy directors, seven different directors of engineering or flight support, five different directors of projects, three different directors of flight operations, numerous 'acting' individuals in many of these positions for extended periods of time, one major and two minor reorganizations, and one consolidation with another center with minor organizational iterations during consolidation."[58]

He further noted significant midlevel management and personnel changes spanning the preceding decade that had affected all of Dryden's projects, and he asked, "How could the board expect any element of the organization to remain stable under these chaotic conditions?"[59]

Personnel assignments in the DAST program had been relatively stable in the 6 months prior to the accident, but Thompson was the only senior manager who had remained in place since the beginning of the program. Changes in the position of DAST project manger were not arbitrary. Two were promoted to larger programs, one retired, and another was given a program assignment for career development purposes. Staffing was bolstered as necessary, such as with the addition of a senior controls analyst and a flight-test engineer. Thompson felt that the board drew its conclusions based on comments from only five people interviewed out of 21 assigned to the DAST program. The board had also apparently failed to interview Dryden senior management personnel or any line supervisors. Thompson complained that the board failed

56. Ibid.
57. Ibid.
58. Ibid.
59. Ibid.

to acknowledge that the DAST program had maintained a nucleus of key personnel for at least 5 years prior to the mishap. "These included the program manager, the project pilot, the operations engineer who is directly responsible for vehicle configuration control and flight vehicle safety, and the lead avionics technician."[60]

Investigators had questioned Dryden's matrix management system in which people with similar skills were pooled for work assignments. Under this system, all engineers may have been assigned to one department (reporting to the chief engineer), but at the same time they may have also been assigned to different projects in which they reported to individual project managers. Thompson's rebuttal noted that although previous Center directors had initially questioned the matrix management philosophy, each had ultimately deemed it the most practical system for Dryden.

Finally, Thomson addressed investigators' criticisms of the priority system for Dryden projects. He pointed out that every organization (from project level to Center management) used a system to prioritize the use of resources, "since no project is either completely self sufficient or has unlimited dollars."[61] He emphasized that priorities often changed in response to a variety of factors, both internal and external. "Almost every Dryden project is either a joint or cooperative project with another NASA center or with another government agency."[62] He added, "The priority system is a pragmatic approach to a complicated scheduling problem and it is by no means a 'sporadic capricious process' as stated by the board."[63]

The DAST Inter-Center Review

As early as December 1978, Dryden Center Director Isaac Gillam had requested that Milt Thompson and chief counsel John C. Mathews investigate management problems associated with the DAST project. This resulted from the project team's failure to meet an October 1978 flight date for the Blue Streak Wing, Langley managers' concerns that Dryden was not properly discharging its project obligations, repeated requests by the project manager for schedule slips, and various other indications that the project was in a general state of confusion. The resulting report indicated that problems had resulted from a

---

60. Ibid.
61. Ibid.
62. Ibid.
63. Ibid.

lack of effective planning at Dryden and had been exacerbated by poor internal communication among project personnel.[64]

Only seven flights were achieved in 10 years. Several were aborted for various reasons and two vehicles crashed, complications that drove up testing costs. Meanwhile, flight experiments with higher-profile, better-funded remotely piloted research vehicles took priority over DAST missions at Dryden. Organizational upheaval also took a toll as Dryden was consolidated with Ames Research Center in 1981 and responsibility for projects was transferred to the Ames Flight Operations Directorate in 1983.

Exceptionally good test data had been obtained through the DAST program, but not in an efficient and timely manner. Despite considering the Firebee/DAST vehicle a quick and reasonably inexpensive option for conducting tasks too hazardous for crewed vehicles, use of off-the-shelf hardware did not guarantee expected results. Just getting the vehicle to fly was far more difficult and far less successful than originally anticipated.[65]

Hardware delays created additional difficulties. The Blue Streak wing was not delivered until mid-1978. The ARW-1 wing arrived in April 1979, a year and a half behind schedule, and it was not flown until 6 months later. Following the loss of the DAST-1 vehicle, the program was delayed nearly 2 years until delivery of the ARW-1R wing. Testing of the ARW-1 was to be followed by tests of Langley's ARW-2 airfoil, but the two DAST accidents threatened further progress.[66]

In the wake of the second major mishap, directors at Ames (which had management responsibility for Dryden at that time) and Langley established a review committee to provide guidance for the future direction of the DAST program. Consisting of four members from Langley and four from Ames-Dryden, and chaired by Langley's Hubert Clark, the committee presented its final report 3 months later, on December 14, 1984.[67]

The committee gave consideration to alternate test vehicles and techniques, but they ultimately concluded that the Firebee drone had been the most appropriate system for the DAST mission. Alternatives, such as testing

---

64. Milton O. Thompson and John C. Mathews, "Investigation of DAST Project," Jan. 22, 1979, DFRC Historical Reference Collection, NASA DFRC, Edwards, CA, file location L1-5-4-15.

65. Loschke et al., "Report of DAST-1R Test Failure Investigation Board."

66. DAST briefing material, Milton O. Thompson collection, April 1987, DFRC Historical Reference Collection, NASA DFRC, Edwards, CA, file location L1-5-4-12.

67. Hubert K. Clark et al., "Report of the DAST Inter-Center Review Committee," NASA Langley Research Center, Hampton, VA, Dec. 14, 1984, DFRC Historical Reference Collection, NASA DFRC, Edwards, CA, file location L1-6-5A-9.

the experimental wing while attached to a larger aircraft or towing the Firebee behind another aircraft, were deemed unacceptable. In order to determine the wing's response to various control techniques, it had to be installed on a free-flying vehicle. Other remotely piloted vehicles, including optionally piloted full-scale aircraft, were dismissed; the Firebee had unique interface and performance capabilities for the DAST program, and changing to a new platform would have negated the already substantial investment in development of the ARW-2 wing, which could be flown only on the Firebee.[68]

Members also considered several options for modifying the Firebee to improve its reliability and increase the probability of successfully completing the DAST program. The committee determined that achieving the desired level of mission reliability would require extensive redesign, fabrication, and testing of vehicle systems. Recommendations included modernization of all electrical/avionics systems, development of improved airspeed controls, and fabrication of a rigid wing for use in the qualification of upgraded vehicle systems. A vertical landing system—a parachute with impact attenuation—seemed the most desirable recovery method.

The committee also recommended establishing a review process to cover all phases of the program, including design, fabrication, and testing. Such a review would be conducted as a joint effort by personnel from Ames-Dryden, Langley, and NASA Headquarters.[69]

In order to mitigate the risk of programmatic delays due to the loss of another vehicle, the committee suggested that two DAST vehicles and two sets of research wings be available. The B-52 was considered the most suitable launch platform, in part because it was subject to NASA control and maintenance practices. There was some concern that due to the bomber's age (it had first flown in 1955), it might become unavailable due to maintenance issues. In fact, committee members had been informed that the aircraft was to be placed into flyable storage in the near future.[70]

It had been anticipated that the ARW-2 flight research program would last 37 months. In view of the many flight programs then under way at Dryden, or planned for the near future, committee members worried that Ames-Dryden managers might be unable to commit the necessary manpower and resources to meet the DAST schedule. Restrictions imposed on another, unrelated project limited flights of a remotely piloted vehicle over populated areas. The committee expressed concern that this philosophy might carry over to other RPV

---

68. Ibid.

69. Ibid.

70. Dryden continued to use the B-52 for numerous projects until it was retired in December 2004.

programs such as the DAST, making flights from Edwards virtually impossible. Ultimately, managers decided to terminate the DAST program altogether, without making any additional flights.[71]

Lessons Learned

The legacy of the DAST program included numerous programmatic and procedural lessons. Notable ones are listed below.

- Mission critical items should be subject to end-to-end systems tests and strict configuration control procedures.
- Whenever possible, system performance should be evaluated in flight at subcritical conditions before being subjected to supercritical conditions.
- Critical aircraft parameters should be available to test monitors in real time.
- Improved presentation and display of warning information would reduce response time by aircraft operators and control room personnel.
- Procedures for the generation and approval of documentation should be consistent and standardized.
- Technical drawings should accurately represent the vehicle configuration, be well organized, up to date, and subject to configuration control.
- Project managers should take positive steps to ensure adequate staffing, training, planning, prioritization, and scheduling discipline.
- A stable nucleus of key personnel should be maintained within the team.
- Even off-the-shelf hardware can present unexpected difficulties.

---

71. Clark et al., "Report of the DAST Inter-Center Review Committee."

The second Perseus-A vehicle, AU-003, is towed aloft from Rogers Dry Lake at Edwards Air Force Base. NASA

CHAPTER 4
Perseus
High-Altitude, Long-Endurance Science Platforms

In the early 1990s, NASA's Earth Science Directorate received a solicitation for research to support the Atmospheric Effects of Aviation project. Because the project entailed assessment of the potential environmental impact of a commercial supersonic transport aircraft, measurements were needed at altitudes around 85,000 feet. Initially, Aurora Flight Sciences of Manassas, VA, proposed developing two remotely piloted research aircraft as part of NASA's Small High-Altitude Science Aircraft (SHASA) program. The fledgling company, founded in 1989 as a follow-on to the Massachusetts Institute of Technology's (MIT's) Daedalus human-powered airplane project, had never attempted to design and build a remotely piloted aircraft. Its first—the Perseus proof-of-concept demonstrator—was tested at Dryden in 1991.

The SHASA effort expanded in 1993 as NASA teamed with industry partners for what became known as the Environmental Research and Sensor Technology (ERAST) project. Goals for the ERAST project focused on the development and demonstration of uncrewed aircraft to perform long-duration airborne science missions. Transfer of ERAST technology to an emerging Unmanned Aerial Vehicles (UAV) industry validated the capability of uncrewed aircraft to carry out operational science missions. NASA's role in the ERAST project was primarily that of facilitator for the development of new technology, drawing together firms interested in building and testing new concepts in this field.

Dryden was responsible for overall management of the ERAST technology demonstration project, with significant contributions by NASA's Ames, Langley, and Glenn Research Centers. Partners in the NASA-industry ERAST Alliance, operated under a Joint Sponsored Research Agreement, included aircraft manufacturers AeroVironment, Aurora Flight Sciences, General Atomics Aeronautical Systems, and Scaled Composites. Other Alliance partners included Thermo Mechanical Systems, Hyperspectral Sciences, and Longitude 122 West. American Technology Alliances served as facilitator for the Alliance, helping team members understand and work toward their common objectives.

Project managers at Dryden were responsible for funding and individual project oversight, establishing overall priorities and technical approaches for meeting project objectives, coordination of facilities for vehicle operations, development and coordination of payloads for demonstration flights, and ensuring that actions taken by ERAST Alliance partners met NASA's future needs for uncrewed vehicles. NASA also conducted independent reviews of vehicle systems and provided input to builders and operators in the interest of improving aircraft design and operations while reducing overall risk.[1]

The ERAST effort progressed with some difficulty due to funding issues, complicated relationships among Alliance partners, and occasional mishaps. Despite these challenges, ERAST researchers set numerous milestones and made extraordinary technological advances before the program was terminated in 2003.

The ERAST effort resulted in a diverse fleet of uncrewed vehicles, several of which were produced by Aurora Flight Sciences. The Perseus-B was powered by a heavily modified, triple turbocharged Rotax engine that ultimately propelled the aircraft to 62,000 feet, the world altitude record for a single-engine, propeller-driven aircraft. Perseus-A, powered by a closed-loop engine that burned gasoline, cryogenically stored oxygen, and recirculated engine exhaust gases, was designed to fly even higher.

Perseus-A

In 1993, Aurora designed and built two Perseus-A vehicles (designated AU-002 and AU-003) with a design goal of carrying a 110-pound payload (later revised to 220 pounds) to an altitude of 82,000 feet while remaining aloft for 5 hours. During test flights, the vehicle carried a 150-pound payload that included equipment for an emergency flight-termination system (FTS). An experimental, closed-system, four-cylinder piston engine recycled exhaust gases and relied on stored liquid oxygen to generate combustion at high altitudes.

Twenty-one flights were conducted over 11 months, all but five flown using AU-003. Because AU-002 had been damaged during shipping, AU-003 was used for the first flight, on December 21, 1993. Eight flights were low-altitude missions to validate mechanical, structural, and flight-control systems. Six medium-altitude flights were conducted to validate closed-loop engine performance and beyond-visual-range flight-control techniques. Finally, there were seven maximum-altitude attempts, although two resulted in premature engine

---

1. "ERAST: Environmental Research and Sensor Technology Fact Sheet," NASA DFRC, Edwards, CA, 2002, *http://www.nasa.gov/centers/dryden/news/FactSheets/FS-020-DFRC.html*, accessed Jan. 23, 2011.

shutdown due to excessive piston temperatures. The remaining five attempts were terminated due to various unrelated mechanical problems. During the 21st and final flight, a gyroscope malfunctioned, resulting in loss of control and structural failure.[2]

Test Operations

Although Perseus team members from Aurora and NASA declared safety to be the test program's highest priority, subsequent events uncovered some deficiencies in this area. Step-by-step procedures were defined and documented for all operations performed on the aircraft before, during, and immediately following each flight. Planners established and reviewed clearly defined test objectives before each flight. Flight cards outlining the desired test points were written by the chief test pilot based on experience and coordination with appropriate personnel. Procedures listed in the test cards ensured safe operation of the aircraft and defined methods for collecting quality test data. In what would later prove a liability, however, formal in-flight emergency procedures had been written only for the takeoff flight phase.

The Perseus flight operations crew consisted of seven people. A flight director was responsible for execution of the flight in accordance with scientific and engineering objectives, established criteria, and procedures. A flight engineer monitored and adjusted aircraft system and payload parameters and performed external communications with controlling agencies. An outside pilot, standing on the lakebed, controlled the aircraft visually during takeoff and landing. An inside pilot controlled the aircraft during flight via instrument displays in a ground control station. A tow truck driver operated a ground vehicle to tow the aircraft to takeoff speed. A safety observer assisted the driver by watching the aircraft during launch, relaying information to the inside pilot, and executing emergency tow-release procedures, if necessary. An aircraft handler performed engine start and prelaunch checks and assisted with launch operations.[3]

The ground control station (GCS) was a modified trailer containing a pilot's control console with a ground-track/location monitor, head-up display showing video from two cameras mounted on the aircraft, and flight instruments

---

2. Michael J. Johnson et al., "Perseus A Final Flight Test Report," Aurora Flight Sciences Corporation, Manassas, VA, 1995, ERAST Project Files FY1994–1998, Jennifer L. Baer-Riedhart personal files.

3. Ibid.

An external pilot controlled the Perseus aircraft visually during takeoff and landing. NASA

displayed on cathode ray tube (CRT) (i.e., television) screens. Two additional CRT screens provided 10 pages of telemetered parameters for use in real-time flight monitoring. During a flight, GCS personnel included the flight director, flight engineer, inside pilot, and backup pilot.[4]

Battling the Wind

Flight 21 with Perseus-A was flown on November 22, 1994, the seventh attempt at a high-altitude mission. The aircraft was loaded with 130 pounds of fuel and 290 pounds of liquid oxygen, giving AU-003 a gross takeoff weight of 1,825 pounds. Test objectives included attaining the maximum design altitude, conducting open-loop airspeed calibration and single-axis autopilot evaluations, and performing minimum-sink-rate airspeed tests during descent. An incident during Flight 19

4. Vance D. Brand et al., "Perseus-A S/N AU-003 Mishap Technical Investigation Report," NASA DFRC, Edwards, CA, 1995, ERAST Project Files FY1994–1998, Jennifer L. Baer-Riedhart personal files.

had alerted engineers to a potential problem with higher-than-indicated power settings. As a result, planners decided to use fuel flow to control engine power.[5]

Flight preparations began before sunrise. There was a 4-hour take-off delay due to surface winds on Rogers Dry Lake, high winds aloft, and the presence of high cirrus clouds. Takeoff approval was finally granted based on an observed trend of decreasing winds and clouds and predictions that high winds aloft would be confined to a narrow altitude band.

Takeoff was uneventful. Per normal procedure, the airplane's two autopilots remained off until sensor parameters (air data, side-slip, and angle of attack) were within reasonable ranges. Once flow-angle vane positions and indicated airspeed reached the appro-

The Perseus ground control station contained simple joystick and switch controls, ground-track/location monitor, head-up display, and flight instruments displayed on CRT screens. NASA

priate values, the longitudinal autopilot was switched on in airspeed-hold mode and the lateral-directional autopilot was switched on in bank-angle-hold mode. The ground pilot flew the aircraft in a racetrack pattern at about 4,300 feet above sea level (2,000 feet above the ground) before climbing through a corridor into the primary operating area.

The airplane encountered turbulence between 3,000 and 5,000 feet mean sea level that resulted in changes of 10 to 12 degrees of bank angle, 3 degrees of sideslip, and 3 degrees of pitch. The pilot described it as the most severe he had encountered on any Perseus flight. As a result of this turbulence, airspeed calibration test points were deleted from the scheduled maneuvers, but a stall test was performed as planned. After recovering from the stall, the pilot trimmed the airplane to climb toward the target altitude of 77,000 feet. The climb rate decreased noticeably at around 26,000 feet.[6]

---

5. Johnson et al., "Perseus A Final Flight Test Report."

6. Brand et al., "Perseus-A S/N AU-003 Mishap Technical Investigation Report."

The Perseus-A had a maneuver speed limit of 61 knots indicated airspeed (KIAS) and a never-to-exceed speed limit of 71 KIAS. NASA

During climb through 30,000 feet, the backup pilot relieved the primary pilot at the controls. Throughout the ascent, the flight engineer performed climb and glide calculations and monitored weather information. His concerns included the unexpected climb rate reduction, indications of a negative ground track, and winds at 36,000 feet that exceeded range safety limits, all of which reduced the likelihood of attaining the desired cruise altitude. While scanning data parameters, he noticed that engine temperature had exceeded prescribed limits. This was sufficient to warrant aborting the mission.

The pilot reduced the throttle setting and began descent procedures that included a normal engine shutdown. As the Perseus-A descended through 34,000 feet, the aircraft experienced slight vertical oscillations that went unnoticed in the control room because personnel were focused on flying the aircraft through a band of high winds and establishing a positive ground speed. Indicated airspeed was increased from 52.5 to 62.4 knots. The Perseus-A had a maneuver speed limit of 61 knots indicated airspeed (KIAS) and a never-to-exceed speed limit of 71 KIAS.[7]

---

7. Ibid.

Members of the recovery crew examine the Perseus-A on the Edwards PIRA. NASA

As the aircraft's oscillations grew in amplitude, everyone in the GCS became aware of the anomalous motions. The pilot reduced airspeed, and the flight director commanded both autopilots be turned off. The flight engineer asked whether he should zero vertical gyro inputs to the autopilots, but the flight director said no. The backup pilot, then acting as pilot-in-command, asked the primary pilot for assistance in turning autopilot switches off.

The heading-hold autopilot was shut off first, followed by the airspeed-hold autopilot. When airspeed suddenly decreased from 59.1 KIAS to 56.1 KIAS, the airspeed-hold autopilot was turned back on. Commanded airspeed immediately increased to 59.2 KIAS.

Diverging oscillatory airspeed and vertical acceleration caused the Perseus-A to pitch up with a corresponding mild upward wing bending, followed by a nose-down pitching motion and wing unloading. The left wing dropped due to an accelerated stall, and at one point, the bank angle reached 80 degrees.[8]

The resulting pullout exceeded aircraft load limits, causing the structural failure of the wing assembly center panel. As both wings departed the vehicle,

---

8. Ibid.

the primary pilot replaced the backup pilot at the control console. Not yet aware that the aircraft had disintegrated, he attempted spin-recovery controls. While the Perseus-A continued to shed parts, a NASA range safety officer (RSO) in the Dryden control room activated the FTS, deploying a parachute that lowered the fuselage remnants to the ground intact. A team of Aurora personnel drove to the impact site, where they deactivated the aircraft's systems, turned off battery power, and vented the liquid-oxygen tank.[9]

Gyro Failure

A joint NASA-Aurora investigation determined that the failure resulted from a malfunctioning pitch gyroscope that fed anomalous attitude signals to the autopilot. The probable cause of the gyro failure was identified as an excessively worn brush assembly in the direct current (DC) spin motor.

The Perseus-A was equipped with multistate feedback controllers in both the longitudinal and lateral-directional axes. The controller on the longitudinal axis used values of angle of attack, airspeed, pitch rate, and pitch attitude, and, based on a reference value, computed an error. The error, multiplied by an altitude-and-airspeed scheduled gain, was fed back to the controller, which then sent appropriate commands to the horizontal control surfaces and/or the throttle actuators.

Because gain in the pitch attitude feedback path was comparatively large, and there were no validity checks or limitations on measured value, an erroneous pitch attitude signal could be fed back to the controller. This had the potential to result in large control-surface rate and deflection commands. Control-surface motion was not limited, allowing maximum surface deflection under certain circumstances. Such conditions left the aircraft open to rate limiting (with associated stability reduction), saturated controls, and the resultant abnormal attitudes and airspeeds. With the autopilot on, stick input from the pilot to command the elevator was insignificant compared to the effect of pitch attitude feedback from the flight-control system.

Investigators concluded that in the Perseus-A mishap, the flight-control system responded to erroneous pitch attitude information from the failing vertical gyro by driving the vehicle outside of its design envelope, ultimately to the point of structural failure.[10]

---

9. Ibid.
10. Ibid.

The Perseus GCS included positions for a pilot and several systems monitors. NASA

Several human factors contributed to the mishap. These were related to deficiencies in the design of the GCS as well as GCS team training.

First, displays in the GCS provided poor situational awareness. This precluded adequate mission monitoring and led to the test team's inability to detect the gyro failure. The control panel featured a very limited set of caution and warning annunciators and no autopilot-mode annunciation. Caution and warning indicators were primarily focused on catastrophic engine failures. None were used to indicate sensor failures. Although telemetry data were available to both the test director and flight-test engineer, the data were presented in a tabular format only, on one of 10 selectable pages. System limits were not indicated by these displays. Operating autopilot modes had to be determined by examination of the autopilot switches. At the time of the mishap, the test director was unable to see these switches due to the GCS layout and was unaware that the airspeed-hold autopilot had been switched back on.[11]

Primary vehicle monitoring was conducted by means of three CRT displays mounted one above the other. Relatively large distances between the three

---

11. Ibid.

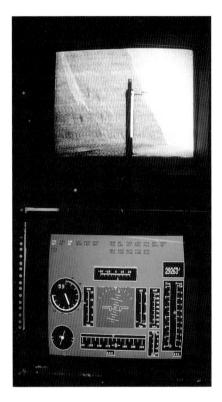

Two screens provided the Perseus pilot with outside visual reference (top) and virtual-cockpit instruments. NASA

screens caused pilots to focus on the head-up display on the center screen, precluding the ability to easily correlate and analyze aircraft flight information. Postflight analysis revealed that a vertical gyro failure could have been diagnosed from the information available on the three CRT displays.

Failure to correlate the displayed information also resulted from the pilot's lack of confidence in the fidelity of the attitude instrument display, which had a relatively slow update rate. During the flight, the instrument-panel display indicated the effects of erratic attitude information from the vertical gyro. That this display did not correlate with the head-up display on the center screen was a clear indication of a failing gyro. The aircraft appeared to stabilize at a nominal wings-level attitude just prior to loss of control, contributing to the pilot's belief that the pitch axis was healthy and leading to his reengagement of the failing pitch (airspeed-hold) autopilot.[12]

Investigators also found that, other than for the takeoff phase, there were no written in-flight emergency procedures for the Perseus-A. It was simply assumed that the GCS team would be able to recognize a problem and have time to determine appropriate corrective action. Further development of emergency procedures had been delayed pending completion of flight tests involving degraded control modes. Program managers assigned these tests a low priority, scheduling them for some time after maximum-altitude flights had been made. Only limited, informal discussion of possible failure scenarios took place, and that was without participation of the entire GCS team. Investigators surmised that the lack of precise terminology for commanding system changes during in-flight emergencies, defined emergency procedures, and training/practice contributed to confusion between the test director and pilots.

---

12. Ibid.

Investigators recommended that the Perseus vehicles not be flown again until the vertical gyro failure was resolved. NASA

According to investigators, GCS test team operations (i.e., elements of test resource management) were degraded in two primary areas. Mission monitoring was degraded due to the entire team's attention being focused on the aircraft's ground track. Team members' attention was not properly divided between aircraft systems and mission conduct monitoring. The lack of adequate caution and warning annunciators led the team to focus on the engine overheat problem and shifted attention from monitoring the aircraft's systems to ensuring that the aircraft's ground track remained within the test range. Although aircraft data indicating motion caused by the failing gyro was available to mission monitors for approximately 8 minutes, the team realized there was an autopilot problem only in the final 45 seconds.[13]

From this point, the failure to establish a safe autopilot configuration resulted primarily from poor communication between the test director and the pilots. The inability to communicate clearly was attributed to three factors. First, the GCS lacked an intercom system, forcing the test director to voice his

---

13. Ibid.

commands in competition with background noise. Second, the pilot was positioned with his back to the test director, having to listen for verbal commands while attempting to fly the aircraft with the displays and controls in front of him. Finally, when the second pilot moved to assist the pilot-in-command in changing autopilot modes, he was confused by the lack of precise commands from the test director. Lacking agreed-upon terminology, he attempted to comply with his interpretation of the test director's intent. He at first correctly turned off both autopilots then mistakenly switched the longitudinal autopilot back on, believing that this was the desired configuration.[14]

The Perseus-A mishap investigation board recommended that the Perseus vehicles not be flown again until the vertical gyro failure was resolved, and a flight readiness review convened to assure overall flight worthiness and safety. The board further recommended inspection of all flight-qualified vertical gyros, reevaluation of the use of brushed motors at high altitude, and possibly obtaining gyros qualified for flight above 40,0000 feet. The members suggested that a method be devised to detect sensor failures in real time. The board recommended that critical GCS displays be relocated within the pilot's scan pattern. Investigators suggested developing well-thought-out emergency procedures and practicing their application with participation by the entire GCS team. The board also recommended improvements to the autopilot that would prevent exceeding parameter limits.[15]

Officials at Aurora Flight Sciences concluded that the basic Perseus-A design was sound. The program set out to develop flexible, affordable robotic aircraft specifically tailored to the needs of the science community, with two prototypes being developed at a cost of about $7 million. An operational cost per flight of roughly $150,000 was comparable to that of scientific-balloon flights with a similar payload weight, and that cost was expected to drop as the technology matured. Although the design altitude was not achieved, the aircraft demonstrated an altitude of 50,000 feet, very close to a record performance for the vehicle class.

Aurora officials recommended implementing a solution to the engine piston temperature problem and upgrading the vertical gyro before resuming test flights. They also suggested that a study be conducted to determine possible subsystem failure modes, as well as development of a reliability upgrade program to identify and resolve any remaining mechanical issues.[16]

14. Ibid.

15. Ibid.

16. Johnson et al., "Perseus A Final Flight Test Report."

The Perseus-B was designed to remain aloft for 24 hours and reach altitudes of around 60,000 feet. NASA

Perseus-B

Aurora engineers designed the Perseus-B to remain aloft for 24 hours. The vehicle was equipped with a triple-turbocharged engine to provide sea-level air pressure up to 60,000 feet. During 2 years beginning with its maiden flight at Dryden, Perseus-B experienced some technical difficulties and several significant incidents.

The first flight of Perseus-B, on October 7, 1994, lasted just 2 minutes and 22 seconds. When sensors detected excessive engine vibrations, the airplane's control system automatically shut the engine down. The inside pilot landed the Perseus-B on Rogers Dry Lake, but the nose gear collapsed during rollout, causing damage to the vehicle's composite nose.[17]

The next two Perseus-B flights occurred in 1996. During the second of these, structural failure of the propeller driveshaft necessitated an emergency return to base. A hard touchdown on the lakebed resulted in major damage: the landing gear tore off, the nose cracked, and the propeller tips were ground off. The vehicle was subsequently transported to Aurora's Manassas facility for repairs.

---

17. Don Nolan, "Research Roundup," *The Dryden X-Press* 36, no. 11 (Nov. 1994).

Nose-gear collapse following the maiden flight of Perseus-B damaged the vehicle's composite nose assembly. NASA

After undergoing a variety of improvements and upgrades—including extending the original 58.5-foot wingspan to 71.5 feet to enhance high-altitude performance—the Perseus-B was returned to Dryden in the spring of 1998 for a series of four flights. Thereafter, additional modifications were made, including the addition of external fuel pods on the wing that more than doubled the fuel capacity to 100 gallons. Engine power was increased by more than 20 percent by boosting turbocharger output. Fuel consumption was reduced with fuel-control modifications and a leaner fuel-air mixture that did not compromise power. After flight operations resumed, the Perseus-B established an unofficial altitude record for a single-engine, propeller-driven, remotely piloted aircraft on June 27, 1998, when it reached an altitude of 60,280 feet.[18]

Test team members experienced some extremely tense moments during a flight on October 1, 1999, after the vehicle suffered an electrical failure, causing loss of control at approximately 49,000 feet. The aircraft had been flying for about 2 hours and 15 minutes when the first indications of voltage fluctuations

---

18. Hallion and Gorn, *On the Frontier: Experimental Flight at NASA Dryden*, pp. 310–311.

During a 1996 flight, a hard touchdown on the lakebed resulted in major damage to the landing gear, nose, and propeller tips. NASA

were detected. The flight was being conducted from Edwards under the guidance of ground-based mission controllers from Aurora Flight Sciences. For safety reasons, the Perseus-B was being flown within restricted airspace over sparsely populated land northwest of the town of Barstow.

As soon as the NASA RSO at Dryden heard over the radio that the vehicle was experiencing control problems, he armed the flight-termination system. One minute later, he executed the terminate command. The arming light extinguished, but the terminate light failed to illuminate. The RSO verified that the signal had been sent and that data indicated the aircraft's engine had shut down. The vehicle, however, remained intact and flying as it descended in a slow spiral that carried the Perseus-B beyond the boundaries of the restricted area. The RSO repeatedly attempted to activate the FTS without success.[19]

The system had, in fact, activated, but the recovery parachute failed to deploy due to a throttle control cable that obstructed the path of the chute-extraction

19. J. Larry Crawford, "Perseus-B Incident Investigation Briefing," NASA DFRC, Edwards, CA, Nov. 5, 1999.

rocket. Ken Cross, who was in the control room, later described the incident as "one the slowest crashes in history."[20] For more than 40 minutes, the Perseus-B continued to descend in a series of spiraling turns over Barstow and surrounding communities as monitors on the ground watched scenes from the airplane's forward-looking camera with a growing sense of panic. "We saw a school, a park, housing tracts, and a strip mall before the video signal was lost."[21]

The aircraft came down in the westbound lanes of Interstate 40 about 4 miles east of Barstow. After touching down on its landing gear, the Perseus-B rolled off the edge of the highway into the dirt median, where the nose gear collapsed. The vehicle sustained moderate damage, primarily to the landing gear and forward fuselage.

Fortunately, traffic was light at the time of the incident. There was no property damage and no fire or injuries on the ground as a result of the crash landing. California Highway Patrol officers secured the scene until the arrival of NASA personnel.[22]

FTS Failure

Dryden flight safety officials formed an accident investigation team to determine the exact cause of the mishap, with assistance from Aurora Flight Sciences operations staff. Through microscopic inspection, mishap investigators discovered a cracked solder joint at a critical junction in a regulator board. Consultation with the part's manufacturer confirmed that such damage would cause voltage transients such as those that had been experienced during the flight. Vibration testing provided conclusive evidence that this was the failure mechanism that initiated the mishap sequence.

Failure of the flight-termination system occurred because the airplane's throttle cable was routed through the path of the parachute-extraction rocket's trajectory. The rocket struck the cable and was deflected aft. With the path of the parachute bag also blocked by the throttle cable, and coupled with aft pull from the rocket, the parachute bag bound against the sides of the container. The FTS failure was thoroughly examined because its functionality was considered a critical public safety issue.[23]

---

20. Ken Cross, personal communication with author, Feb. 9, 2012.

21. Ibid.

22. Leslie Williams, "Perseus-B Damaged in Crash on California Highway," NASA Press Release 99-115 (Oct. 1, 1999).

23. Crawford, "Perseus-B Incident Investigation Briefing."

Failure of the flight-termination system led to an unplanned landing of Perseus-B on a major highway. NASA

The RSO observed that inadequate display information resulted in uncertainty as to whether the terminate signal had been transmitted. There was no attempt to select a backup transmitter or alternate control panel. At the time, RSO procedures did not address FTS failures. Although FTS status information was available to the RSO, a design deficiency resulted in loss of power to the telemetry system upon receipt of the terminate command.

The incident also raised an issue with regard to emergency procedures and preparation. Emergency response was delayed because the pilot never declared an in-flight emergency. Otherwise, the pilot remained calm throughout the incident and did everything possible to recover the aircraft.

Investigators recommended that Dryden's range safety office formalize oversight and approval authority for the design and testing of all FTS system components used in Dryden aircraft, including those maintained and operated by tenants. The investigation board further recommended that NASA inspectors play a greater role in inspecting flight-critical systems of remotely piloted aircraft prior to flight.[24]

Continued use of parachute recovery systems was deemed acceptable, provided that an independent FTS capable of placing the vehicle on a ballistic

24. Ibid.

trajectory was installed. The board suggested that all flight-termination systems should meet a minimum reliability standard of 0.999 at 95-percent confidence.

Board members felt that FTS telemetry should provide the RSO with a positive indication of all commanded functions and should remain active throughout the entire flight. They also recommended that all parachute recovery systems be fully tested, including a complete flight-configuration-installed end-to-end systems test. Investigators further recommended that preflight procedures be developed for all vehicles equipped with a parachute FTS to ensure that the deployment path remained clear of obstacles.[25]

At the time of the Perseus-B mishap, Dryden had range safety responsibility but did not have flight safety responsibility for this type of aircraft. Flight safety was left to the operating contractor (in this case, Aurora), and NASA oversight of project activity was kept to a minimum. The investigation board recommended that steps be taken to increase the involvement of NASA personnel to include engineering oversight, pre-mission and day-of-flight inspections, and monitoring of critical data in the control room.

In order to improve confidence in FTS activation, investigators recommended that after two attempts with no indication that the vehicle was indeed terminated, the RSO should switch to a backup system and retransmit the signal. This, however, would have made no difference in the Perseus-B incident due to the obstruction that prevented deployment of the recovery parachute.

There had been numerous instances during the mission when the downlink system experienced data dropouts. Investigators recognized that due to the amount of equipment associated with tracking, acquisition, distribution, processing, and display of data, it was often difficult to locate the source of such dropouts. The only way to determine whether the problem originated with the aircraft or ground equipment was to record the automatic gain control (AGC) voltages from telemetry receivers used at the tracking station. The board recommended that the range make it standard practice to record AGC voltages during all missions involving remotely piloted vehicles.[26]

Lessons Learned

The Perseus program suffered a number of incidents resulting from mechanical malfunctions and human factors. It may be worth remembering that this was Aurora's first attempt to build and operate a full-scale remotely piloted aircraft.

---

25. Ibid.
26. Ibid.

Additionally, as a "Host Program," flight safety was the contractor's responsibility. NASA provided no direct support for quality assurance.

- Human factors in the Perseus-A mishap included GCS design deficiencies and personnel training.
- GCS displays should be designed to provide maximum situational awareness with system status indicators and warning annunciators that are prominent, clear, and easy to read.
- Displayed information should be updated in real time and easy to correlate.
- Develop in-flight emergency procedures prior to flight and have written checklists available for GCS personnel.
- Develop precise terminology for use during nominal operations and in-flight emergencies, define emergency procedures, and subject all relevant personnel to rigorous contingency training.
- Ensure that test controllers, monitors, and pilots have clear and adequate communication channels.
- Thoroughly inspect critical safety systems to ensure proper function. The Perseus-B FTS functioned properly, but it was inhibited by the airplane's throttle cable—an airframe/system integration problem.
- Provide a backup FTS capability.
- The host agency should take an active role in regard to safety, engineering oversight, inspection, and monitoring.

The Theseus was a high-altitude, long-endurance vehicle capable of carrying a 700-pound payload to 60,000 feet. NASA

CHAPTER 5

Theseus

Mission to Planet Earth

Aurora Flight Sciences also built a larger vehicle named Theseus that was funded by NASA through the Mission To Planet Earth environmental observation program. Aurora and its partners, West Virginia University and West Virginia's Fairmont State University, built the Theseus for NASA under an innovative $4.9 million fixed-price contract. The Mission to Planet Earth project office was managed by NASA's Goddard Space Flight Center. Dryden hosted the Theseus flight-test program, providing hangar space and range safety. Aurora personnel were responsible for flight testing, vehicle flight safety, and operation of the aircraft.

The twin-engine, unpiloted vehicle had a 140-foot wingspan and was constructed primarily from composite materials. Powered by two 80-horsepower, turbocharged piston engines that drove twin 9-foot-diameter propellers, it was designed to fly autonomously at high altitudes, with takeoff and landing under the active control of a ground-based pilot.[1]

With the potential to carry 700 pounds of science instruments to altitudes above 60,000 feet for durations of greater than 24 hours, the Theseus was intended to support research in areas such as stratospheric ozone depletion and the atmospheric effects of future high-speed civil transport aircraft engines. The long-endurance vehicle was to be used for independent validation and verification of measurements taken from space-based platforms such as the Agency's Earth Observing System family of satellites. The Theseus aircraft was envisioned as a key element of a system that would provide approximately 5,000 hours per year of data collection and correlative measurements.

Preliminary design work for the Theseus prototype was complete in June 1994 and followed by 22 months of fabrication, construction, static testing, and systems integration. For ease of transport, the vehicle was designed for

---

1. NASA DFRC, "Aurora's Theseus Remotely Piloted Aircraft Crashes," News Release 96-63, Nov. 12, 1996, *http://www.nasa.gov/centers/dryden/news/NewsReleases/1996/96-63.html*, accessed June 11, 2011.

Initial flights of the Theseus demonstrated basic handling qualities at low and medium altitudes. NASA

quick assembly and disassembly, and it was designed to fit inside standard-size shipping containers. Major components included the fuselage, engines, nacelles, wing center section, left and right midspan wing panels, and outboard wing panels. The center panel was joined to each engine nacelle and to the fuselage with a four-bolt joint at each location. The horizontal tail and vertical tail were also removable.[2]

Into the Air

Aurora Flight Sciences shipped the Theseus prototype to Dryden in the spring of 1996 for flight testing. Following extensive ground testing, the aircraft was declared flight-ready on May 21, but high winds delayed the maiden flight for several days.

Piloted by Einar Enevoldson (under contract to Aurora) from the ground control station, the Theseus took to the air for the first time on May 24, 1996. Project manager and flight director Tom Clancy noticed an oscillation in the propeller pitch speed controller as the Theseus accelerated for takeoff. Clancy called for a mission abort, but as Enevoldson pulled back on the throttles, reducing the downward pitching moment caused by a high-thrust centerline, the airplane lifted off and began to climb. It reached an altitude of about 70 feet before losing airspeed and dropping. Enevoldson applied power and managed to stabilize the vehicle at about 15-feet altitude and initiated a 300-foot-per-minute climb. After experiencing more problems with propeller oscillations, he brought the craft down for a safe landing on the lakebed.[3]

Aurora engineers analyzed design information and flight-test data to identify the causes of the prop-speed oscillation. Contributing factors included modeling errors, software development delays, and actuator circuitry limits. The results of modifications and testing of various subsystems were incorporated into simulations and used to design a new prop-speed controller. High-speed taxi tests validated the simulation models. There were no oscillation problems

2. Matthew G. Hutchison, Matthew T. Velazquez, and David W. Vos, "Flight Testing of the Theseus Prototype," presented at the Association for Unmanned Vehicle Systems International, 244th Annual Symposium and Exhibition, Baltimore, MD, June 1997.
3. Karen Palmer, "Theseus Flies—Ahead of Schedule," *The Aurora Data Link* 12, second quarter (1996).

during the second flight, on June 28, when the Theseus team completed low-altitude test objectives.[4]

The test team conducted several additional checkout flights for envelope expansion with the goal of carrying proof-of-concept science payloads to design altitude. Objectives included a demonstration of the vehicle's autonomous navigation capabilities, precision landings, continuous flight for 8 hours, carriage of a science payload to an altitude of at least 50,000 feet, and determination of the aircraft's operational ceiling. [5]

The sixth flight—a medium-altitude performance evaluation—occurred on November 12, 1996. The flight plan called for a climb to 20,000 feet followed by a series of maneuvers. Weather conditions were excellent. All objectives were met, and the Theseus began its descent for landing. Enevoldson, the ground pilot, was flying the aircraft in a heading-hold autopilot mode. During a maximum-rate left turn made to keep the aircraft within designated airspace boundaries, he attempted to switch the autopilot to bank-hold mode to prevent turn reversal due to heading command "wraparound." He inadvertently turned the switch too far, turning the autopilot off.

The aircraft quickly exceeded its 15-degree autopilot bank limit, reaching a maximum of 30-degrees bank. Enevoldson began recovering from the maneuver in open-loop mode then reengaged the autopilot. One-tenth of a second later, the right wing came off and the airplane spiraled toward the ground.[6]

For such contingencies, the Theseus was equipped with an independent flight-termination system, including two rocket-deployed parachutes designed to allow recovery of the entire structure intact in the event of a catastrophic accident. Despite repeated commands, transmitted by the range safety officer to activate the system, the parachutes failed to deploy. As the stricken aircraft plunged toward the ground, increasing air loads caused it to further disintegrate. The Theseus crashed near the north end of Rogers Dry Lake and within the boundaries of Edwards Air Force Base. There were no injuries to ground personnel or property damage beyond the loss of the aircraft.[7]

---

4. Martin Gomez, "Theseus completes low-altitude test objectives in successful second flight," *The Aurora Data Link* 13, supplemental (July 1996).

5. Karen Palmer, "NASA launches Theseus envelope expansion program," *The Aurora Data Link* 14, third quarter (1996).

6. Matt Hutchison, "Theseus Prototype: Results from Flight Testing and Mishap Investigation," presented at NASA DFRC, Edwards, CA, July 22, 1997, ERAST Project Files FY1994–1998, Jennifer L. Baer-Riedhart personal files.

7. John S. Langford, "Theseus Loss, Plans Detailed," *The Aurora Data Link* 15, third quarter (1997).

Investigators examine Theseus wreckage at the crash site near the north end of Rogers Dry Lake, within the boundaries of Edwards Air Force Base. NASA

Through the Labyrinth

Aurora named a five-member accident investigation board made up of company and NASA officials. With assistance from four NASA Centers and the Theseus Project Office, investigators examined the wreckage, interviewed all participants, and reviewed flight telemetry records and engineering documentation. Of primary interest were the cause of the wing's structural failure and the reason why the emergency parachute system failed to deploy.

Following a lengthy investigation, the board concluded that all systems aboard the Theseus aircraft and its associated ground control station were functioning normally at the time of the accident. Investigators found no anomalies in the flight-control system, propulsion system, communications equipment, or other subsystems. The board discounted the possibility that clear-air turbulence may have been a factor but could not rule out the possibility of a progressive failure such as cracking due to ground, flight, or test loads prior to the accident.[8]

The in-flight structural failure was found to have occurred at a joint between two wing panels. Separation of an end closeout rib from the wing skin eliminated the structural-load path between the aft spar and a connecting pin that was part of the joint. The failure then propagated forward as the rib continued to peel away from the wing skin. The outer wing panel rotated forward and upward, crushing the leading edge and inducing torsion loads into the main spar joint, causing it to fail. Investigators believed that the initial failure occurred at loads well within the aircraft's design envelope.[9]

The accident resulted from a combination of several contributing factors. First, a design error resulted in the underestimation of loads carried in the outer panel's end rib as it transferred loads from the aft spar to the connecting pin. An aircraft with a high-aspect-ratio wing, such as that on the Theseus, generates significant chordwise aerodynamic loads. These are typically tension loads located in the vicinity of the aft wing spar. At the interface between the two wing panels on Theseus, the continuity of the rear spar was broken and reestablished through the wing joint. According to NASA Langley aerospace engineer Juan R. Cruz, loads from the rear spar were transferred across the wing joint through the end ribs of each wing panel and then to the end ribs through metal pins. The end ribs had to sustain loads that were trying to pull them off their respective wing panels. Cruz noted that routing the rear spar loads through the end ribs was a serious design flaw. "A preferable design involves connecting the rear spars of the two wing panels directly, without first transferring the loads through the end ribs."[10]

Investigators also discovered a manufacturing error. The designers had intended the end rib to be manufactured as a single piece, but due to an improperly drawn template guide, it was instead manufactured in two pieces. Although the error was detected and reinforcing plies added to the joint between the two pieces, no calculations were made to analyze the strength of the modified joint.

---

8. Ibid.

9. Ibid.

10. Juan R. Cruz, Structural Mechanics Branch, NASA Langley Flight Research Center, Hampton, VA, letter to Jeff E. Bauer, NASA DFRC, July 30, 1997, ERAST Project Files FY1994–1998, Jeff Bauer personal files.

Analysis of the adhesive used to assemble the wing revealed that it had a much lower strength than had been assumed during the design process. In particular, the adhesive used in a secondary bond had much lower peel strength than allowable due to a combination of factors, including application and curing procedures. The wing structures had been designed in such a way that secondary-bond joints were never supposed to be placed in peel, but the affected bondline was placed in peel due to the previously noted design and manufacturing errors.[11]

Cold temperatures further degraded the adhesive's peel strength. The flight took place in mid-November, a time of typically cold weather at Edwards. Additionally, it was the first time the Theseus aircraft had attained a 20,000-foot altitude. At that height, the airplane was exposed to temperatures of around −4 degrees Fahrenheit (−20 degrees Celsius).

Structural testing of the Theseus airframe during development concentrated on large out-of-plane bending loads through the wing joint and on determining structural aeroelasticity modes. No proof tests were conducted to evaluate the wings' capability to carry in-plane bending loads.

Engineers determined that dynamic loads due to autopilot transients during the banking turn were outside normal operational parameters but within limits for the airplane's structural design envelope. The pilot's actions—inadvertently turning off the autopilot—triggered the accident but were not the cause of the failure.[12]

Investigators also examined the failure of the flight-termination system. To alleviate program risk exacerbated by a relatively small budget and short schedule, the Theseus had been equipped with a system that would ensure that the aircraft could not depart controlled airspace, would enhance the safety of persons and property on the ground, and would allow recovery of the aircraft in the event of catastrophic failure.

Upon activation, the FTS was designed to shut down both engines and fire the two rocket-deployed parachutes. The two-step activation process included transmitting the "arm" command for one full second, followed by a one-second "terminate" command. Transmission records showed that the "arm" command was sent for just 0.84 seconds—insufficient time to arm the system. As a result, the terminate command was not processed by the airplane's onboard systems. Because the system was never armed, repeated attempts to send the terminate

---

11. Langford, "Theseus Loss, Plans Detailed."
12. Ibid.

The Theseus FTS panel was designed to allow the RSO to shut down the airplane's engines and fire two rocket-deployed parachutes in the event of an emergency. NASA

command had no effect. Investigators concluded that operator error was the single factor leading to the FTS failure.[13]

Because the FTS radio link and activation system were Government-furnished equipment, NASA conducted an independent investigation into the FTS failure. The results were in agreement with those of the Aurora investigation board. NASA investigators concluded that since the aircraft broke up within the boundaries of the military reservation, range safety was never

---

13. Ibid.

compromised. Testing of recovered components indicated that proper 1-second command duration would have armed the FTS. Although the right-wing FTS antenna was lost when the aircraft disintegrated, slightly degrading the radio link, telemetry data showed that there was still sufficient link margin available at the time that the arm command was sent. NASA investigators determined that while the range safety officer relied on verbal communications to execute FTS duties, protocols for verbal FTS activation were neither properly documented nor adhered to.[14]

Recommendations for addressing the FTS problem included performing an end-to-end review of FTS procedures and hardware. Investigators suggested providing the RSO with direct physical control of FTS activation, control panels with timing indicators, and downlink telemetry data indicating FTS status.[15]

Slaying the Minotaur

Aurora's investigation board produced several key recommendations pertaining to the aircraft's structure, flight-control system, and flight-termination system. These included redesigning the wing joint and using an adhesive with improved low-temperature peel characteristics. The board recommended making changes in Aurora's engineering processes to increase the level of review in design and analysis. Investigators also suggested that a method of nondestructive inspection should be developed for use with secondary structural bond-lines. In order to prevent inadvertent shutoff, the board recommended that autopilot switches should be modified to prevent accidental disengagement. Aurora investigators agreed with NASA recommendations that the FTS arming command should provide feedback to the operator, but they also suggested that the flightcrew should have the ability to activate the FTS.[16]

The Aurora investigation concentrated on specific technical causes of the accident, but it did not address the underlying management and programmatic decisions that allowed the situation to occur. In order to effectively absorb and implement lessons from the Theseus mishap, Aurora officials ordered a company-wide standdown to review and reflect upon its causes. Subsequently, the company's internal management policies were reviewed, critiqued, and

14. Ronald Young, "Theseus Flight Termination System Performance Investigation," briefing presented at NASA DFRC, Edwards, CA, Jan. 21, 1997, ERAST Project Files FY1994–1998, Jeff Bauer personal files.

15. Ibid.

16. Langford, "Theseus Loss, Plans Detailed."

extensively revised. Aurora created a new office to oversee quality assurance and expanded the company's interaction with NASA engineers at Langley in reviewing design requirements, practices, test procedures, and quality assurance methods.[17]

Despite not having achieved the program's goals, flight testing of the Theseus prototype successfully demonstrated key aspects of technologies for a high-altitude, long-endurance research platform. Although the aircraft attained a maximum altitude of only 20,000 feet during the test series, simulations based on collected data indicated that a 50,000-foot altitude could have been achieved. A production variant of the Theseus, with improvements, might have been capable of reaching 61,000 feet with a 30-hour endurance and transcontinental range.[18]

Although the loss of the prototype was a setback, Aurora officials hoped to use the lessons learned in the development of a second, similar aircraft. Parts had been manufactured for a second Theseus airframe, but funding for the project was not immediately forthcoming. John Langford, president of Aurora Flight Sciences, noted that, "Losses are expected in this kind of testing."[19]

During the course of the aggressive, low-cost program, the possible loss of a prototype vehicle was considered an acceptable and appropriate risk. The Theseus flight-test program was conducted in such a way that the mishap did not result in injury, loss of life, or damage to property. The accident was not caused by any failure fundamentally related to the vehicle concept. Data collected and lessons learned from the flight-test program and the accident investigation laid the essential foundation for the development of similar future aircraft. Addressing the issue of risk in these types of projects during a March 1997 speech, NASA Administrator Daniel S. Goldin said, "We'll fly them, and we're going to crash them, and we'll learn."[20]

Lessons Learned

The technical causes of the accident were more easily solved than the underlying management and programmatic decisions that set the stage for the mishap. Once again, human factors were a significant contributor.

17. Ibid.
18. Hutchison, Velazquez, and Vos, "Flight Testing of the Theseus Prototype."
19. David Colker, "Pilotless Aircraft Destroyed During Test Flight," *Los Angeles Times*, Nov. 13, 1996.
20. Hutchison, "Theseus Prototype: Results from Flight Testing and Mishap Investigation."

- Engineering processes should include in-depth review of vehicle design and analytical methods.
- Nondestructive inspection techniques can identify structural flaws and weaknesses prior to flight.
- Autopilot controls should be designed to prevent inadvertent disengagement.
- In some situations, it may be desirable for the flightcrew to have the ability to activate the FTS.
- Government and contractor engineers should work together to review design requirements, practices, procedures, and quality assurance methods.

The Helios Prototype in flight, demonstrating normal dihedral conditions. NASA

CHAPTER 6

Helios Prototype

Chasing the Sun

Innovative designers at AeroVironment, in Monrovia, CA, took a markedly different approach to the ERAST challenge. Engineers at the company, which was founded by Paul MacCready, designer of the human-powered Gossamer Albatross and solar-powered Gossamer Condor, sought to create an airplane that would climb above 50,000 feet using solar-electric propulsion. Their design concept featured a simple flying wing with a rectangular planform spanning 98.5 feet with an 8-foot chord and weighing around 500 pounds. The AeroVironment team did not have to create a brand new vehicle for the ERAST program. A proven platform was already available.

Dark Origins

In 1981, AeroVironment obtained funding from a Government agency for a classified demonstration of a concept for a high-altitude, solar-powered, uncrewed aircraft. Called HALSOL, it was designed to fly day and night at high altitudes (above 65,000 feet) for long periods in temperate and tropical latitudes, too far from either pole to capitalize on seasonal 24-hour sunlight. As a result, the power-generating photovoltaic panels would have to collect at least twice the solar energy needed for daytime flight and store the surplus for use at night.

Compared with aircraft driven by internal-combustion engines, solar-powered aircraft require a large wing area relative to aircraft weight (i.e., with low wing-loading). Moreover, the higher the desired operating altitude, the more power is required to sustain flight. Therefore, the wing area covered by photovoltaic cells had to be maximized to collect sufficient sunlight. Conditions at the desired operating altitudes for HALSOL required wings too large for a conventional cantilever design in which the wing is projected from a large central mass, the fuselage. Additionally, it had to be possible to increase the solar-collection area without forcing an increase in the thickness of the spar (the main structural element of the wing). The HALSOL configuration was described as a span-loader

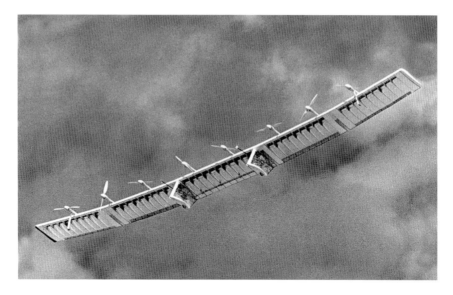

Since the HALSOL was not equipped with autonomous flight controls or solar cells, all flights were conducted using radio control and battery power. AeroVironment

because weight was distributed as evenly as possible across the wingspan. This configuration was characterized by low weight and high structural strength. In case of an emergency, HALSOL was equipped with a parachute recovery system that was programmed to deploy upon the loss of communications link.[1]

The vehicle's wing, built in five segments of equal span, featured a carbon-fiber-composite spar and Styrofoam ribs braced with spruce and Kevlar, covered with thin Mylar plastic film. Two gondolas hung from the center segment, each designed to carry a payload, radio control and telemetry electronics, and other gear. The gondolas also provided the landing gear. Each gondola had dual baby-buggy wheels in front and a bicycle wheel in back. HALSOL was propelled by eight electric motors driving variable-pitch propellers. There were two motors on the center wing segment, two on each inner wing segment, and one on each outer wing segment. The aircraft's total gross weight was about 410 pounds, including a 40-pound payload.[2]

Nine HALSOL test flights took place between June and July 1983 at a remote desert test site. Since the aircraft had not yet been fitted with autonomous flight controls or solar cells, all flights were conducted using radio control

---

1. Nicholas J. Colella and Gordon S. Wenneker, "Pathfinder and the Development of Solar Rechargeable Aircraft," *Energy & Technology Review* (July 1994).

2. Ibid.

and battery power. The ground pilot initiated takeoff from a dry lakebed while observers followed the slow-moving craft on bicycles. During one flight, the batteries in the pilot's control box went dead. Loss of link initiated automatic deployment of the recovery chute. The aircraft descended in a gentle spiral and landed on the dry lakebed, sustaining no damage.[3]

The test series validated the HALSOL vehicle's aerodynamic and structural properties, mechanical performance, and flight characteristics. Researchers concluded, however, that technologies for efficient photovoltaic cell and energy storage technology were not yet sufficiently mature. The HALSOL craft was put into storage for nearly 10 years before being resurrected for a Ballistic Missile Defense Organization (BMDO) project.

In 1993, BMDO officials selected the HALSOL vehicle as one of two platforms for Responsive Aircraft Program for Theater Operations (RAPTOR), an effort to create a system for defending against tactical ballistic missiles. The concept of operations included a UAV called Theater Application Launch-On-Notice (TALON) that would have been armed with antimissile weapons and another UAV equipped with sensors to detect missile launches. The latter, designated Pathfinder, required long-duration, high-altitude capabilities. Subsequently, the HALSOL vehicle was renamed RAPTOR/Pathfinder and equipped with high-efficiency, lightweight solar cells as well as more-efficient electric motors and propellers.[4]

Out of the Black, Into the Blue

With the addition of solar panels to the upper surface of the Pathfinder wing, five low-altitude checkout flights were made under the BMDO program at Dryden in the fall of 1993 and early 1994. These tests demonstrated flight using a combination of solar and battery power. Although budgetary and political considerations eventually contributed to the demise of the RAPTOR program in 1994, the Pathfinder prototype was not retired. In 1995 it was transferred to the joint NASA-industry ERAST program, where it underwent further modifications. AeroVironment technicians removed the two center engines, added a complete set of solar panels, and configured the aircraft

---

3. NASA, Ken Cross, "UAV Mishap Summary Report," AS&M System Safety, NASA DFRC, Edwards, CA, Nov. 2003, Ken Cross personal files.

4. "Unmanned Aerial Vehicles (UAV) Program Plan," Defense Airborne Reconnaissance Office, Washington, DC, April 1994.

with a 50-pound-payload capability. The unusual craft attained an altitude of 50,500 feet, a record for solar-powered aircraft.[5]

After additional upgrades and checkout flights at Dryden, ERAST team members transported the Pathfinder to the U.S. Navy's Pacific Missile Range Facility (PMRF) at Barking Sands, Kaua'i, HI, in April 1997. Predictable weather patterns, abundant sunlight, available airspace and radio frequencies, and the diversity of terrestrial and coastal ecosystems for validating scientific imaging applications made Kaua'i an optimum location for testing. During one of seven high-altitude flights from the PMRF, the Pathfinder reached a world altitude record for propeller-driven as well as solar-powered aircraft, at 71,530 feet.[6]

In 1998, AeroVironment technicians modified the vehicle to include two additional engines and extended the wingspan from 98 feet to 121 feet. Renamed Pathfinder Plus, the craft had more efficient silicon solar cells developed by SunPower Corp., of Sunnyvale, CA, that were capable of converting almost 19 percent of the solar energy they received to useful electrical energy to power the motors, avionics, and communication systems. Maximum potential power was boosted from about 7,500 watts on the original configuration to about 12,500 watts, allowing the Pathfinder Plus to reach a record altitude of 80,201 feet during another series of developmental test flights at the PMRF.

NASA research teams, coordinated by the Ames Research Center and including researchers from the University of Hawaii and the University of California, used the Pathfinder/Pathfinder Plus vehicle to carry a variety of scientific sensors. Experiments included the detection of forest nutrient status, observation of forest regrowth following hurricane damage, measurement of sediment and algae concentrations in coastal waters, and assessment of coral reef health. Several flights demonstrated the practical utility of using high-flying, remotely piloted, environmentally friendly solar aircraft for commercial purposes. Two flights funded by a Japanese communications consortium and AeroVironment emphasized the vehicle's potential as a platform for telecommunications relay services. A NASA-sponsored demonstration employed remote-imaging techniques for use in optimizing coffee harvests.[7]

AeroVironment engineers ultimately hoped to produce an autonomous aircraft capable of flying at altitudes of around 100,000 feet for weeks, or even months, at a time through the use of regenerative fuel cells. Building on their

---

5. "NASA Dryden Fact Sheet—Pathfinder Solar-Powered Aircraft," NASA Web site, 2001, *http://www.nasa.gov/centers/dryden/news/FactSheets/FS-034-DFRC.html*, accessed June 13, 2011.

6. Ibid.

7. Ibid.

Pathfinder (1981-1997)

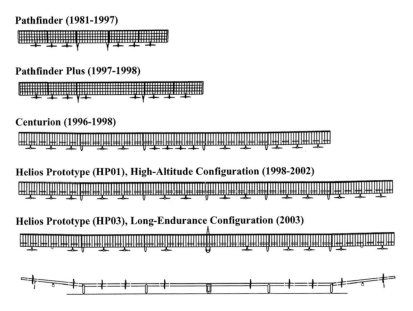

Pathfinder Plus (1997-1998)

Centurion (1996-1998)

Helios Prototype (HP01), High-Altitude Configuration (1998-2002)

Helios Prototype (HP03), Long-Endurance Configuration (2003)

AeroVironment developed a family of solar-powered craft with incrementally increased performance capabilities. NASA

experience with the Pathfinder/Pathfinder Plus, they subsequently developed the 206-foot-span Centurion.

Plans originally called for the construction of two airframes. The first, Centurion, would be used for demonstrating high-altitude capability. The second, to be called Helios, was to be flown on a 96-hour long-duration mission at an altitude of 50,000 feet. Key technologies developed for Pathfinder/Pathfinder Plus were improved for use on the Centurion and Helios vehicles.

Three test flights at Dryden in 1998, using only battery power to drive 14 propellers, demonstrated the Centurion's handling qualities, performance, and structural integrity. During its final flight, the Centurion carried a simulated payload weighing 605 pounds. Bill Parks of AeroVironment, Centurion's chief engineer and flight-test director, noted that the high-gross-weight payload was a major objective of the flight-test program: "We verified the performance of the aircraft while flying in a high gross-weight configuration. We came here with a new variant of our proven platform and it performed exactly as expected; there were no aircraft systems that had to be corrected. It doesn't get any better than that."[8]

---

8. SunPower Corporation, "Centurion Solar Aircraft Finishes Flight Tests Early-Moving Closer to a Commercial Satellite Substitute," SunPower Press Release, Dec. 3, 1998 *http://us.sunpowercorp. com/about/newsroom/press-releases/?rellD=179501*, accessed June 13, 2011.

Budget constraints in early 1999 forced NASA and AeroVironment to develop plans to attain both ERAST goals of altitude and endurance using only a single airframe. Consequently, plans to build the Helios vehicle were scrapped. Instead, the Centurion was modified to include a 247-foot wingspan, strengthened center wing panels, and a fifth landing gear pod. The 4 center motors were redistributed along the new center wing panels, but the total remained at 14. The modified vehicle, with a performance goal of 100,000-feet altitude and 96-hour mission duration, was renamed Helios Prototype. As with its predecessors, a ground pilot remotely controlled the craft, either from a mobile control van or a fixed ground station. The aircraft was equipped with a flight-termination system—required on remotely piloted aircraft flown in military restricted airspace—that included a parachute system plus a homing beacon to aid in determining the aircraft's location.[9]

Flight-test plans were based on an incremental approach that allowed researchers to slowly build up to a demonstration of the aircraft's design goals. Initial testing of the Helios Prototype, then known as HP99, included six battery-powered flights for evaluation of handling qualities, stability and control, response to turbulence, and the use of differential motor thrust to control pitch. Researchers used the opportunity to check out and calibrate instrumentation required for the planned solar-powered high-altitude and long-endurance flights. Four of these flights were used to assess the high-altitude configuration and two, with the aircraft ballasted to simulate inclusion of the planned regenerative fuel cell system (RFCS) hardware and solar array, were conducted to assess the performance of the heavier long-endurance configuration.[10]

Chasing the Sun

In 2000, technicians at AeroVironment began upgrading the Helios Prototype to its HP01 high-altitude configuration, adding new avionics, environmental control systems, and a SunPower solar array. Following installation of more than 62,000 solar cells, the aircraft was transported to the PMRF for high-altitude test flights. On August 13, 2001, the HP01 reached an altitude of 96,863 feet, a world record for sustained horizontal flight by a winged aircraft.

---

9. "Helios Prototype: The forerunner of 21st century solar-powered 'atmospheric satellites,'" NASA Web site, 2002, *http://www.nasa.gov/centers/dryden/news/FactSheets/FS-068-DFRC.html*, accessed June 13, 2011.

10. Thomas E. Noll et al., "Investigation of the Helios Prototype Aircraft Mishap," Volume I, Mishap Report, January 2004, *http://www.nasa.gov/pdf/64317main_helios.pdf*, accessed June 13, 2011.

By this time, a full-scale prototype of the RFCS pod had been constructed, but the hydrogen-oxygen fuel cells and electrolyzers were not yet sufficiently reliable for flight testing. Due to schedule and budget constraints, AeroVironment proposed switching to a consumable primary fuel cell system (PFCS) derived from existing technology used by the automotive industry. Analysts had determined that this would allow flight tests to continue as planned and provide the Helios Prototype with a 7-to-14-day flight-duration capability. These factors were important, as AeroVironment officials wished to attract other commercial and Department of Defense customers and bring the vehicle's high-altitude, long-endurance capability to market as soon as possible. The final test series was scheduled for 2003, the final year of the ERAST program. Without the possibility of schedule or budget relief, the desire to accomplish a major milestone before program termination drove the decision to switch to a PFCS. This reduced some of the technical risks but made it harder for the team to consider other risk reduction efforts such as a low-altitude test flight at Dryden prior to the first high-altitude, long-duration mission.[11]

In December 2001, technicians began modifying the Helios Prototype to its long-duration mission configuration, known as HP03. The center landing gear pod was replaced with a fuel cell pod weighing approximately 520 pounds. Two high-pressure hydrogen fuel tanks weighing approximately 165 pounds each (including 15 pounds of liquid hydrogen), and associated plumbing, were added beneath the outer wing panels at motor-pylon locations nos. 2 and 13. Four motors (nos. 2, 6, 9, and 13) were removed, leaving just 10 motors to power the aircraft. For weight reduction, a wing spar made from an aluminum tube was replaced with one made from carbon-fiber-composite material. Technicians installed new propellers, optimized for flight at 65,000 feet. Wingtip panel incidence was reduced from 1.0 degree to zero degrees. The forward row of solar cells on the center wing panels and the first two rows from midwing and wingtip panels were removed. Technicians also removed servos from the wingtip panels and fixed the wingtip elevators at –2.5 degrees (trailing edge up). Due to the highly flexible nature of the wing, landing gear was installed on the wingtip panels. Engineers revised flight-control-system autopilot gains and programmed gain scheduling with respect to airspeed. Three battery packs were reconfigured and installed in pods nos. 2 and 4 to mass-balance the aircraft. By the end of 2002, the PFCS had been designed and fabricated. In April 2003, it was integrated into the HP03, and technicians completed a series of combined systems tests. The HP03 had a gross weight

---

11. Ibid.

of 2,320 pounds, an increase of 735 pounds from that of the HP01 during its altitude-record flight in 2001.[12]

The aircraft's load-carrying structure was constructed mostly of lightweight composite materials. The main wing spar, made of carbon fiber, was thicker on top and bottom to absorb bending loads during flight. It was wrapped with Nomex and Kevlar to provide additional strength. The wing ribs were made of epoxy and carbon fiber. The leading edge of the wing consisted of aerodynamically shaped Styrofoam, and the entire wing was wrapped in a thin, transparent plastic skin. The airfoil was not tapered or swept, having an 8-foot chord (aspect ratio of 31) with a maximum thickness of 11.5 inches (constant from wingtip to wingtip) and 72 trailing-edge elevators spanning the entire wing. The main landing gear and battery power system were enclosed within aerodynamically shaped underwing pods attached at each wing panel joint.

The HP03 aircraft was powered by 10 brushless direct-current electric motors rated at 2.0 horsepower, or 1.5 kilowatt each. Each motor was equipped with a two-bladed propeller, 79 inches in diameter, made of composite materials. Steering during flight was accomplished by increasing power to the four outboard motors on one side while decreasing power to the four on the opposite side. For pitch control, a computer sent commands to servomotors to actuate the elevators. To provide adequate lateral stability, engineers had designed the outer wing panels with a 10-degree dihedral (upsweep). To prevent wingtip stall during low-speed maneuvers and landings, the wingtips had a slight upward twist.[13]

The primary objective of the HP03 flight-test series was the successful demonstration of a hydrogen-air fuel cell to sustain flight overnight at 50,000 feet. Starting in February 2002 and continuing through January 2003, a series of design and technical reviews were carried out to thoroughly examine aircraft configuration changes, structural loads, stability and control, and aeroelastic models and predictions. The results led to a decision to strengthen the wingtip spars so that their structural margins would be consistent with those along the wing spar under design load conditions.[14]

Although the structural, stability and control, and aeroelastic safety margins were more reduced on the HP03 than they had been on the HP01 configuration, NASA and AeroVironment engineers felt that they were sufficient for conducting the long-endurance demonstration. Additionally, the mass distribution with the PFCS was significantly different than that of the initially proposed configuration with the RFCS. Equipped with the RFCS, the aircraft

---

12. Ibid.
13. Ibid.
14. Ibid.

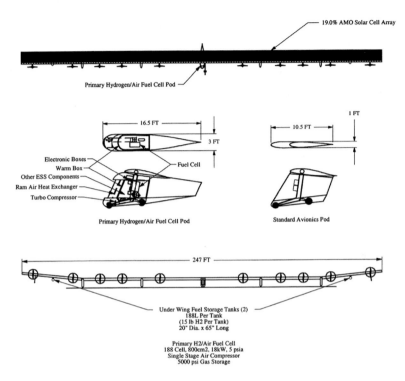

Addition of a center-mounted 520-pound fuel cell and two 165-pound fuel tanks beneath the outer wing panels changed the Helios Prototype from a span-loader to a point-loaded configuration. NASA

would have required only two regenerative fuel cell pods located about one-third the distance from aircraft centerline to the wingtips. Installation of three pods for the PFCS configuration resulted in increased point loading.

On May 15, 2003, the HP03 was flown for the first time in a short hop just 2 feet above the runway for 10 seconds to verify the proper wing dihedral distribution. Additionally, the exercise provided the team with an opportunity to conduct a dress rehearsal of all necessary preflight assembly and test procedures. A postflight assessment indicated that the aircraft had approximately the correct dihedral distribution and that all aircraft systems, fuel cell pod, and ground support equipment were working well, with the exception of the solar array, which suffered two broken bus bars. Using test data from this brief flight, engineers fine-tuned the aircraft's mass distribution, wingtip panel incidence angle, elevator settings, and flight-control-system gains to establish a safe operating envelope for high-altitude missions.[15]

---

15. Ibid.

The first true test flight occurred several weeks later, on June 7, to validate the vehicle's handling characteristics and aeroelastic stability with the fuel cell system and gaseous hydrogen storage tanks installed. The Helios team demonstrated readiness of the vehicle's systems, fuel cells, flight controls, flight support equipment, range support instrumentation, and procedures required for conducting a long-duration flight. Data telemetered in real time validated the predicted aeroelastic characteristics and demonstrated that the HP03 was aeroelastically stable at the flight conditions expected for the long-endurance flight demonstration. Although the HP03 was estimated to be capable of an approximately 30-hour flight duration at 50,000 feet altitude, the test had to be aborted 15 hours after takeoff due to leaks in the coolant system as well as in compressed-air lines feeding the PFCS. As a result of this leakage, the test team was unable to activate the fuel cell system.

Turbulence levels and wind during this flight were uncharacteristically light. Engineers worried that airspeed variations during turns, high sideslip at low-power/low-altitude conditions, and sensitivity of wing dihedral to power setting over the entire flight envelope might make the aircraft more difficult to handle under typical weather conditions in the test area. In order to address these concerns, the team modified the aircraft. Propeller pitch was flattened from −5.5 degrees to −8.0 degrees. Power throttle scaling on the two outboard motors was reduced, and the drag mode was eliminated. Technicians increased the flight-control-system autopilot longitudinal gains and the ratio of the airspeed-hold gain to the pitch-attitude-damping gain. Wingtip panel incidence was increased from zero to 0.5 degrees.[16]

The second flight was scheduled for June 26, 2003. Objectives included clearing the aircraft's flight envelope for the new configuration changes and for the 50,000-to-60,000-foot-climb/glide performance needed for the planned long-duration mission. Researchers wanted to verify stable operation of the fuel cell and compressor system and to achieve a rated power of 18.5 kilowatts at 50,000 feet. It was also important to run the fuel cell system for at least 2 hours to develop confidence that it would be able to run all night, and to demonstrate a rapid shutdown of the fuel cell pod and night restart on battery power. Finally, researchers wanted to develop a fuel cell performance sensitivity matrix that could be used to optimize performance for the long-duration mission.

---

16. Ibid.

During a flight with excessively turbulent conditions, the Helios Prototype's wing structure experienced extreme dihedral. NASA

Excessive Dihedral and Pitch Rate

By 5 a.m. on June 26, crews had already spent more than 5 hours preparing the Helios Prototype for flight and readying the stationary and mobile ground control stations. All aircraft and fuel cell systems performed well during preflight testing. Weather predictions indicated that light winds were expected from the east. Upper level cirrus clouds had moved over the Hawaiian Islands overnight but were forecasted to be out of the area by takeoff time.

As the HP03 was being towed to the runway, AeroVironment and NASA personnel held a go/no-go review. Based on the results of preflight tests, compliance with mission rules, and the weather forecast, the team was "go" for flight. Although weather constraints were acceptable by mission rules, the meteorologist indicated that conditions were marginal due to the close proximity of a shear line that could create turbulence during climb to altitude.[17]

Takeoff was delayed 90 minutes due to a change in wind direction, necessitating moving the aircraft to the opposite end of the runway, and upper-level cirrus clouds blocking sunlight to the aircraft's solar panels. Additionally, the meteorologist advised that light to moderate turbulence was expected, not merely possible.

When the HP03 finally took off, winds were blowing at 7 knots and scattered cumulus clouds were shadowing parts of the runway. As a result, the vehicle's climb rate was slightly less than normal. The stationary crew spent the first 10 minutes helping the mobile pilot navigate around the clouds. About 3 minutes later, the stationary pilot noted that the aircraft was apparently encountering turbulence.

Observers watched the upward bowing of both wingtips, a normal phenomenon that increased the flexible wing structure's dihedral. Over the next several minutes, however, the wing's dihedral increased beyond the normal range on two occasions; in both cases, this was accompanied by a mild pitch oscillation. Each time, the wing dihedral quickly returned to normal and the oscillations damped out. Since these events occurred while the mobile crew was transferring control to the stationary crew, neither was aware of either the high dihedral or the pitch oscillations.[18]

On several occasions, the pilot of a chase helicopter, observing whitecaps in the ocean, advised turning the HP03 to avoid wind shear and find smoother air. The stationary pilot complied while attempting to stay as close as possible to the planned flightpath.

The stationary pilot selected a visual display from the wingtip video camera because he believed it would provide the best indication of wing dihedral and the aircraft's response to turbulence. The HP03 was cruising at an altitude of approximately 2,800 feet when it began experiencing airspeed excursions of about ±2 feet per second. Based on the wingtip video, observers felt that the dihedral seemed high for the indicated airspeed (38 feet per second). Normally, the camera view would have looked across the wingspan. Instead, the camera was pointed toward the top of the center wing panel. This high dihedral condition persisted, and the aircraft began a series of pitch oscillations of increasing intensity.[19]

---

17. Ibid.
18. Ibid.
19. Ibid.

The pilot conferred with the stability and control engineer, confirming that the proper procedure for reducing the dihedral was to increase airspeed. When the pilot increased the aircraft's speed by an additional foot per second, the dihedral decreased slightly then increased beyond 30 feet. Airspeed fluctuations indicated that the HP03 was experiencing large pitching motions and diverging airspeed excursions of about ±10 feet per second. As a result of the persistent high dihedral, the aircraft became unstable in a highly divergent pitch mode. The pilot noted that he thought the aircraft was in a large phugoid oscillation because the airspeed excursions were almost off the scale, the amplitude of the unstable pitching motion nearly doubling with every cycle.

In an attempt to stop the pitching motion, the pilot initiated emergency procedures and immediately turned off the Airspeed Hold switch. At this point, the vehicle was already pitching down sharply and accelerating to approximately two-and-a-half times the maximum design airspeed. At these extreme conditions, aerodynamic loads shattered the foam sections of the leading edge of the right wing panel near the hydrogen fuel tank. Solar cells and skin began peeling off the upper surface of the wing. As progressive failure of secondary structure continued, the HP03 disintegrated and fell into the ocean. Elapsed time from the first effort to diagnose and correct the high wing dihedral condition to the start of structural breakup was just 91 seconds.[20]

Helios Mishap Investigation

Salvage teams recovered most of the largest pieces—about 75 percent of the vehicle by weight—approximately 10 miles off the coast of Kaua'i. Heavy items such as the fuel cell system sank in mile-deep water and were not recovered.

Investigators determined that the mishap resulted from the inability to predict, using available analysis methods, the aircraft's increased sensitivity to atmospheric disturbances, such as turbulence, following vehicle configuration changes required for the long-duration flight demonstration.[21]

At takeoff, environmental conditions appeared to be within acceptable parameters. The island served as a windbreak over the airfield and for some distance offshore. This "wind shadow" was bounded to the north, south, and above by zones of wind shear and turbulence. Compared with previous flights, the vehicle's longer exposure to the island's leeside turbulence and lower

---

20. Ibid.
21. NASA, "Helios Mishap Report Released," NASA DFRC, Edwards, CA, Sept. 3, 2004, NASA press release from Jeff Bauer personal files.

The Helios Prototype sheds parts as it plummets toward the ocean. The FTS worked exactly as designed. NASA

shear-line penetration coincided with the airplane's sensitivity to turbulence, possibly compounded by the narrow corridor between shear lines as observed from the chase helicopter.

Mission planners sought to avoid turbulence by having the aircraft climb as rapidly as possible to altitudes characterized by smoother air. Because the HP03 flew at a somewhat higher airspeed than previous solar-powered configurations without increasing climb rate, the airplane was exposed to greater turbulence at lower altitudes for a longer period. About 30 minutes into the flight, turbulence caused the aircraft to develop an unexpected, persistent high-dihedral condition that led to instability in pitch. During the final seconds of flight, the vehicle exceeded its design airspeed, and resulting dynamic pressures caused it to disintegrate.

Investigators found two root causes of the mishap. First, a lack of adequate analysis methods led to an inaccurate risk assessment of the effects of vehicle configuration changes. This resulted in an inappropriate decision to fly the HP03 in a configuration that was highly sensitive to turbulence. Second, configuration changes, driven by programmatic and technological constraints, altered the aircraft from a span-loader to a highly point-loaded mass distribution without changing the basic vehicle structure. This reduced the structure's load-bearing capabilities as well as the margin of safety for flight operations in turbulent air.[22]

Engineering analyses performed prior to the HP03 flights had accurately predicted the wing dihedral shape only under smooth air conditions. Although data showed that the vehicle would be unstable with a wing dihedral greater than 30 feet, engineers did not predict the degree to which the aircraft would be sensitive to disturbances, the inability of the aircraft's structure to restore itself to a more-nominal dihedral following disturbance, or the highly divergent nature of the vehicle's unstable pitch mode. During the mishap flight, the first encounter with turbulence did not result in the development of high-dihedral conditions. The next two, occurring in the span of 3 minutes, caused the HP03 to develop a dihedral of about 30 feet and pitch oscillations that damped out on their own without pilot interaction. Neither the mobile nor the stationary flightcrew interpreted associated airspeed variations as periodic oscillations, but rather as typical aircraft response to turbulence. The final encounter with turbulent conditions resulted in a dihedral approaching 40 feet accompanied by rapidly divergent pitch oscillations.

Investigators noted that the strong dihedral response was surprising compared to events of the prior test flight. The first HP03 mission had been flown under unusually benign wind conditions, and pitch oscillations encountered during earlier test flights of Pathfinder/Pathfinder Plus, Centurion, and Helios were generally mild. Flightcrews had time to deliberate on a course of corrective action, and any pitch oscillations encountered quickly damped out when such action was taken. Lack of a history of large, sustained dihedral deflections may have instilled false confidence in predicted pitch-instability parameters.[23]

The unusual structural configuration of the Helios Prototype presented researchers with a nonlinear stability and control problem involving interactions among the flexible composite airframe, unsteady aerodynamics, flight-control system, propulsion system, environmental conditions, and flight dynamics. Analytical tools available at the time failed to provide engineers with sufficient understanding of the way these factors interacted with the vehicle's stability and control characteristics. As a result, the Helios Prototype mishap investigation board made several recommendations.

First, the board recommended the development of more-advanced, multi-disciplinary—structures, aeroelasticity, aerodynamics, atmospherics, materials, propulsion, controls, etc.—time-domain analysis (a method of representing a

---

22. Noll et al., "Investigation of the Helios Prototype Aircraft Mishap."

23. Ibid.

waveform by plotting amplitude over time) techniques appropriate to highly flexible air vehicles. Second, the board suggested that ground-test procedures appropriate to this class of vehicle should be developed to validate new analytical methods and predictive techniques. The board also recommended the development of multidisciplinary modeling techniques capable of describing the nonlinear dynamic behavior of aircraft modifications. In order to improve project management, the board suggested that for highly complex research programs, the use of expertise from all NASA Centers could improve technical insight. Finally, the board emphasized the need to provide adequate resources to future programs to allow for more incremental testing when major configuration changes significantly deviate from the initial design concept.[24]

During the course of their investigation, the seven members of the board discovered that the AeroVironment-NASA technical team had developed most of the world's existing knowledge base regarding design, development, and testing of high-altitude, long-endurance aircraft. Board President Thomas E. Noll wrote, "This class of vehicle is orders of magnitude more complex than it appears,"[25] noting that team members from AeroVironment and NASA had identified and solved the toughest technical problems.

Lessons Learned

The Helios Prototype was a very unusual aircraft that presented numerous technical and operational challenges for the test team. Lack of accurate predictive modeling created a hidden vulnerability under turbulent atmospheric conditions, especially after the span-loader was changed to a point-loaded configuration.

- Configuration changes can significantly affect aircraft behavior, making accurate predictive modeling essential to increasing the margin of safety.
- Lack of adequate analytical methods can lead to inaccurate risk assessment.
- Ground-test procedures appropriate to the class of vehicle can be used to validate analytical methods and predictive techniques.
- Program managers should draw on a wide range of expertise and use a multidisciplinary approach for developing accurate analytical models.

---

24. Ibid.
25. Ibid.

- Adequate resources should be provided to allow for incremental testing when configuration changes significantly deviate from the initial design concept.

In order to boost the X-43A to the required Mach number and altitude for scramjet ignition, the vehicle was attached to the front of a modified Pegasus launch vehicle. This artist's concept shows the mated stack prior to separation. NASA

CHAPTER 7
Hyper-X
Hypersonic Air-Breathing Propulsion

In 1996, NASA initiated an ambitious program to advance research in high-speed air-breathing propulsion technologies from laboratory experiments to flight test. The multiyear effort, called Hyper-X, was aimed at conducting flight research in the hypersonic (Mach 5 plus, or more than 3,600 miles per hour [mph]) speed regime. The first vehicle in a proposed series of high-risk, high-payoff endeavors was designated X-43A. Researchers hoped to achieve target speeds of Mach 7.0 to 10.0, the fastest ever attained by an air-breathing aircraft. The fully autonomous X-43A was designed to serve as a test bed for a supersonic combustion ramjet (i.e., scramjet) propulsion system.

As in all jet engines, scramjets provide thrust by igniting fuel in compressed air and exhausting expanding gases to propel the aircraft forward. Most jet airplanes, capable of speeds in the subsonic to Mach 2.2 (1,600 mph) range, use turbojet or turbofan engines that have rotating blades to compress the air. Ramjets are theoretically capable of propelling aircraft to Mach 6 (4,600 mph) by using the plane's forward motion alone to bring air into the combustion chamber. But the vehicle must first be boosted to approximately Mach 3, and air entering the engine inlet must be slowed to subsonic speed for ignition regardless of the aircraft's speed. In a scramjet engine, airflow through the inlet and combustion chamber remains supersonic, a feat that NASA engineers compare to "lighting a match in a hurricane."[1]

Micro Craft, Inc., based in Tullahoma, TN, received a contract to manufacture four X-43A vehicles, each designed for one-time use. Other members of the contractor team included Boeing, GASL Corporation, and Accurate Automation, Inc. The program was managed for NASA at the Langley Research Center, while the Dryden Flight Research Center provided flight-test facilities and personnel.[2]

---

1. Guy Gugliotta, "With 'Scramjet,' NASA Shoots for Mach 10," *The Washington Post*, Nov. 10, 2004.
2. Jay Miller, *The X-Planes—X-1 to X-45* (Hinckley, UK: Midland Publishing, 2001).

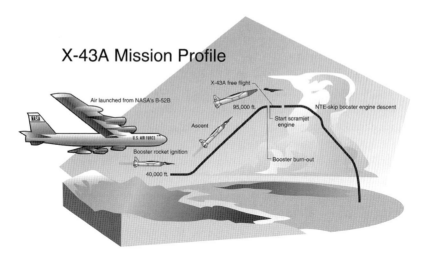

Each Hyper-X vehicle was designed for single use, returning all data via telemetry before falling into the Pacific Ocean. NASA

The Technical Challenge

Researchers initially planned four X-43A flights at incrementally increased speeds from Mach 5 to Mach 10. Technical problems with both airframe and propulsion system development resulted in several lengthy delays of the first flight, originally planned for 1998. In fact, the first X-43A was not delivered to Dryden until October 1999. As a result of funding constraints, acquisition of one of the four vehicles was cancelled, as was the Mach 5 flight. The revised flight schedule called for a Mach 7 flight at 95,000 feet and two to Mach 10 at 110,000 to 120,000 feet.[3]

In order to boost the X-43A to the required Mach number and altitude for scramjet ignition, the vehicle was attached to the front of a modified Pegasus launch vehicle called the Hyper-X Launch Vehicle (HXLV). The combination of X-43A research vehicle, adapter, and HXLV was called the stack. In order to achieve test conditions suitable for collecting the desired hypersonic data, engineers calculated a launch trajectory that necessitated operating the booster at a lower altitude and higher dynamic pressure than those of a typical Pegasus trajectory.[4]

In order to meet the demands of the new flight profile, engineers had to make several modifications to the Pegasus first stage. These included changes to

---

3. Ibid.

4. Robert W. Hughes et al., "Report of Findings: X-43A Mishap," NASA, Sept. 2002, *http://www. nasa.gov/pdf/47414main_x43A_mishap.pdf*, accessed June 3, 2011.

the booster's three stabilizer fins to protect against increased aerodynamic heating; additional thermal protection altered the shape of the fins and increased their leading-edge radius. The booster's wings also received increased thermal shielding. Additionally, allowances had to be made for in-flight deformation of the HXLV due to increased structural loading as compared to that experienced in a standard Pegasus mission profile.[5]

The Hyper-X concept of operations called for the HXLV-X43A stack to be carried aloft by NASA's modified B-52 and released at an altitude of around 20,000 feet above the Pacific Ocean. The HXLV was designed to boost the X-43A to stage-separation conditions, which were selected to achieve the desired Mach number and dynamic pressure for the powered portion of each test. This boost phase included a 2.5-g pull-up followed by a –1.5-g pushover. Following rocket-motor burnout at an altitude of approximately 95,000 feet and speeds approaching Mach 7, the X-43A was programmed to separate from the booster and autonomously stabilize its flightpath to achieve test conditions. A cowl would then open, allowing airflow through the inlet duct. The X-43A's airframe and engine was actually an integrated scramjet in that the forward section of the fuselage served as the inlet, channeling air to the engine, while the fuselage section aft of the engine functioned like a rocket nozzle. Silane (hypergolic fuel) was then injected into the flow path, and hydrogen fuel was added once the flame ignited.[6]

After ignition, the scramjet would operate for approximately 7 seconds. Following engine shutdown, the cowl would close and the remaining 6 minutes of flight would be devoted to collecting aerodynamic data during high-speed gliding maneuvers. The X-43A was not equipped with recovery systems or landing gear; instead, it would simply fall into the ocean within the confines of the Western Test Range, more than 500 miles from its launch point off the California coast. All data were collected through telemetry data linked to ground receiving stations during flight.[7]

The original Pegasus booster was designed to place small payloads into low Earth orbit. In contrast, the HXLV was required to remain at lower altitudes for the duration of its engine burn. After dropping away from the B-52, its rocket motor ignited, propelling the stack horizontally for up to 13 seconds before a pull-up maneuver sent the vehicle into a steep climb. Instead of lofting its payload into space, toward the end of its mission the HXLV performed a

---

5. Curtis Peebles, *Road to Mach 10: Lessons Learned from the X-43A Flight Research Program* (Reston, VA: AIAA, 2008).

6. Gugliotta, "With 'Scramjet,' NASA Shoots for Mach 10."

7. Miller, *The X-Planes–X-1 to X-45*.

The X-43A-HXLV stack drops away from the B-52 over the Pacific Missile Range. NASA

pushdown maneuver, entering a slightly negative angle of attack at a specified altitude in order to increase velocity. By the time the rocket motor burned out, the stack was again flying in a near horizontal attitude.

The stack's transonic aerodynamics posed significant challenges for the NASA engineers because the shape of the X-43A interacted with the control surfaces of the HXLV in unpredictable ways. Wind tunnel tests were conducted to analyze these effects, but engineers approached the problem as if the HXLV were an off-the-shelf item. The existing database of Pegasus wind tunnel results seemed to confirm that the HXLV flight profile could be successfully executed. In fact, Hyper-X engineers were much more concerned with issues involved in separating the X-43A from the stack at hypersonic speeds while the vehicles were subjected to complex flow conditions and high dynamic pressures. By comparison, the launch phase was considered to be a known quantity.[8]

---

8. Peebles, *Road to Mach 10*.

The X-43A accelerates following booster ignition. The stack's transonic aerodynamics posed significant engineering challenges because the shape of the X-43A interacted with the control surfaces of the HXLV in unpredictable ways. NASA

A Dream Deferred

The first Mach 7 test was scheduled for June 2, 2001, over the Naval Air Warfare Weapons Sea Range west of Point Mugu, CA. The B-52 carrying the X-43A stack took off from Edwards and was crewed by Dana Purifoy and Frank Batteas. Brian Minnick served as launch panel operator for the HXLV, and Matt Redifer was assigned as X-43A panel operator. Two NASA F-18 chase planes accompanied the flight. One carried a videographer, the other a still photographer for engineering documentation.[9]

The captive-carry portion of the flight was nominal except for an alternator that failed prior to takeoff. One hour and fifteen minutes later, Purifoy and Batteas had the B-52 on course at the planned launch altitude of just over 20,000 feet. At the command "Launch, launch, launch," the stack dropped away from the pylon. Five seconds later, the rocket motor ignited. This was accompanied by a regulator malfunction in the HXRV adapter that allowed venting of gaseous nitrogen, but the anomaly had no impact on the events to follow.

Ten seconds after release, HXLV steering guidance engaged as planned, followed by a pitch-up maneuver. A rudder and two fins, mounted equidistantly at the aft end of the booster, provided steering and stabilization. At the start of pitch-up, the stack began experiencing a divergent roll oscillation that increased in frequency. When the rudder actuator reached its electrical current limit, it ceased responding to commands, indicating a rudder actuator stall. Moments later, the starboard fin departed the vehicle, quickly followed by the

---

9. Ibid.

port fin, rudder, wing, and the X-43A. Telemetry was soon lost, and the range safety officer activated the flight-termination system.[10]

Although the booster incorporated elements and a basic design configuration from the Pegasus, the HXLV was launched and flown in an environment that differed significantly from previous Pegasus experience. At the time of failure, the HXLV was flying at 22,244 feet, whereas a standard Pegasus vehicle would have been operating at 40,000 feet for the same flight duration. The HXLV experienced dynamic pressure of 650 pounds per square foot (psf) at Mach 1.0, more than twice that typically experienced by the Pegasus for the same Mach number. This increase in dynamic pressure at transonic conditions proved to be a significant factor in the mishap.[11]

Into the Deep

NASA's Associate Administrator for Aerospace Technology appointed a mishap investigation board with representatives from the Marshall Space Flight Center, Langley Research Center, Goddard Space Flight Center, Kennedy Space Center, Dryden Flight Research Center, and NASA Headquarters. Robert W. Hughes, from Marshall, served as chairman of the 10-person investigative team, supplemented by independent experts from NASA Centers and representatives of Orbital Sciences and Micro Craft.

Unlike most accident investigations, this one was challenged by the fact that there was no wreckage available for study. All physical evidence from the research mission had fallen into the Pacific Ocean in approximately 1,200 feet of water, and no attempt was made to recover the flight hardware. Investigators instead used recordings of telemetered data and applied a fault tree–based investigation methodology. This choice was based on the complexity of the vehicle's physical and functional systems, the multiorganizational character of the Hyper-X team, availability of fault trees used in risk assessments during program development, and board members' familiarity with the fault tree investigation process. Initially, investigators focused on assessments of the physical, functional, engineering, and operational characteristics of the Hyper-X program in relation to telemetered data from the flight. Of 613 faults that were examined, 11 were determined to be direct contributors to the mishap, and three were deemed potential contributors.[12]

---

10. Hughes et al., "Report of Findings: X-43A Mishap."
11. Ibid.
12. Ibid.

Divergent roll oscillatory motion caused by excessive control-system gain resulted from inadequate analytical modeling that allowed engineers to overestimate system safety margins. NASA

As a root cause, investigators determined that the HXLV failed because the vehicle's control-system design was deficient for the trajectory flown and could not maintain vehicle stability during transonic flight. This resulted from inadequate analytical modeling that allowed engineers to overestimate system safety margins. The key phenomenon that triggered the failure was divergent roll oscillatory motion at a frequency of 2.5 hertz (Hz), primarily caused by excessive control-system gain. Rudder actuator stall, a consequence of these oscillations, accelerated the loss of control. Neither phenomenon had been predicted in preflight analyses.[13]

Modeling inaccuracies with regard to the fin actuation system and aerodynamics, as well as insufficient variation of modeling parameters (parametric uncertainty analysis), were significant contributors to the mishap. Investigators found inaccuracies in both Pegasus heritage and HXLV-specific models. Fin actuation system inaccuracies resulted from discrepancies in modeling electronic and mechanical components as well as underprediction of the fin actuation system compliance used in analytical models. Aerodynamic modeling inaccuracies stemmed from errors in the incorporation of wind tunnel data into mathematical models and failure to model changes in the vehicle's outer mold line associated with the thermal protection system.[14]

---

13. Ibid.
14. Ibid.

Less significant contributing factors included errors detected in modeling mass properties. Investigators also found potential contributing factors in the areas of dynamic aerodynamics and aeroservoelasticity.

Examples of insufficient parameter uncertainty analysis were discovered in regard to aerodynamics, fin actuation systems, and control systems. Based on the results of the investigation, Hyper-X engineers recalculated linear stability predictions using corrected models and computed stability gain margins for all axes. Revised predictions proved that the nominal model was still insufficient to predict vehicle loss of control and that parameter uncertainty had to be included. Recalculation of the X-43A flight conditions, while accounting for parameter uncertainties in the analyses, allowed the engineers to replicate the mishap conditions. This result was confirmed with nonlinear time history predictions using a six-degree-of-freedom flight dynamics simulation of the X-43A/HXLV stack. Investigators determined that no single factor or potential contributing factor was to blame. Conditions leading to the flight mishap could only be reproduced when all of the modeling inaccuracies, with uncertainty variations, were incorporated in a nonlinear simulation model.[15]

Return to Flight

Following release of investigation results in March 2002, the Hyper-X team pushed ahead with plans for a second flight. Investigators had recommended an extensive review of hardware, software, and systems; improved wind tunnel tests and computational fluid dynamics analysis; and refinement of mathematical models used to create the flight plan. This included running full end-to-end mission simulations at both Langley and Dryden. Parameter uncertainties were introduced during Monte Carlo analyses that, even with conservative margins, indicated a 97-percent probability of successful flight through X-43A separation.

Management and review procedures were also revised to strengthen the existing integrated product team (IPT) structure through improved communications and teamwork. The IPT for each discipline was responsible for developing a return-to-flight plan detailing necessary action items, and all actions carried out by each IPT were subject to approval by an engineering review board. Documentation ensured that each action item was tracked until closed.[16]

---

15. Ibid.

16. Peebles, *Road to Mach 10*.

In order to reduce dynamic pressure on the HXLV, the stack had to be launched from a higher altitude. The only way to achieve this was through the reduction of booster weight by removing some of the solid propellant, a hazardous task. Only two HXLV boosters remained available for the X-43A flight program, and no funds were available for a replacement. If engineers required a static test firing of the modified "Hyper-X Lite" booster, it would mean sacrificing plans for the proposed Mach 10 flight of the third X-43A. Ultimately, the engineering team decided it could accurately predict the rocket's performance with the reduced propellant load.

Nearly 2 years and 9 months after the mishap, and following modifications to the HXLV booster's motor and fin actuation system, improvements to the launch pylon, additional modeling and simulation, and systems tests, the second X-43A was declared ready for flight. A successful launch on March 27, 2004, was followed by the separation of the X-43A from the stack and scramjet ignition. The X-43A achieved a peak velocity of Mach 6.83, making it the fastest air-breathing aircraft in history to that point.

In the wake of this success, the third and final X-43A flight was scheduled for November 16, 2004. This time, the vehicle achieved a record speed of Mach 9.68, just shy of the Mach 10 target. Despite the problems encountered early in the program, Hyper-X turned out to be a resounding success.[17]

Lessons Learned

The X-43A mishap resulted from inaccurate predictive modeling with regard to the launch vehicle.

- Inadequate analytical models allowed engineers to overestimate system safety margins.
- Based on the inaccurate model, the vehicle's control system was deficient for flight conditions at the planned trajectory.
- Insufficient variation of modeling parameters contributed to the mishap.
- No single factor or potential contributing factor was to blame. Conditions leading to the flight mishap could only be reproduced when all of the modeling inaccuracies, with uncertainty variations, were incorporated in a nonlinear simulation model.

---

17. Ibid.

An RQ-1L Predator UAV flies over the Nevada Test and Training Range. USAF

CHAPTER 8

Manning the Unmanned

Humans in the Loop

Increasing use of remotely piloted aircraft for military and civil applications has drawn attention to their accompanying high mishap rate. In fact, some studies have shown accident rates several orders of magnitude greater than those associated with crewed aircraft. By far, the greater percentage of RPA mishaps is attributed to human factors.[1]

Use of the term "unmanned" to describe any sort of autonomous or remotely piloted aircraft is somewhat disingenuous because it is often misunderstood to mean that there is little or no human-systems integration involved. In fact, RPA operations involve numerous people at every level. For any RPA to function requires human-systems interface in every aspect of control and maintenance. In most respects, RPA operations are identical to those involving conventionally crewed aircraft. The fact that the RPA operator is geographically separated from the aircraft necessitates particular attention to human-factors engineering when developing such systems.

In a 2004 report for the U.S. Department of Transportation, researchers reviewed and analyzed RPA accident data to identify human-factors issues. Data collected from the U.S. Army, Navy, and Air Force indicated that human factors contributed to between 21 percent and 68 percent of RPA mishaps. Since RPA systems vary widely in size and complexity, the specific percentage of human-factors involvement varied according to aircraft model. Researchers also found that many of these mishaps could have been anticipated through the analysis of user interfaces and procedures for each system.[2]

---

1. Anthony P. Tvaryanas, William T. Thompson, and Stephan H. Constable, "U.S. Military Unmanned Aerial Vehicle Mishaps: Assessment of the Role of Human Factors Using Human Factors Analysis and Classification System," HSW-PE-BR-TR-2005-0001, 311th Performance Enhancement Directorate, 2005, *http://www.wpafb.af.mil/shared/media/document/AFD-090226-154.pdf*, accessed July 27, 2011.
2. Kevin W. Williams, "A Summary of Unmanned Aircraft Accident/Incident DATA: Human Factors Implications," DOT/FAA/AM-04/24, U.S. Department of Transportation, Office of Aerospace Medicine, *http://www.hf.faa.gov/docs/508/docs/uavFY04Mishaprpt.pdf*, accessed July 27, 2011.

Human error can almost always be traced to a variety of underlying causes. These may include workload, fatigue, poor situational awareness, inadequate training or supervision, lack of proper crew coordination, and poor design. Any of these, alone or in combination, can degrade human performance and increase the likelihood of a mishap.[3]

Cognitive Factors: Pilot Workload

Pilot workload results from a combination of task demands and human response to these demands. Ranging from simple radio-controlled models to full-scale aircraft with varying degrees of autonomy, RPA produce a range of workload levels. Vehicles used for long-duration missions can present operators with extended periods of low workload (navigation and monitoring system health) interspersed with periods of high workload (takeoff, landing, and operation of sensors or other mission equipment). Operator response to malfunctions or unexpected severe weather conditions can result in brief periods of intensely high mental workload.[4]

The demands of even basic maneuvers can place a considerable workload on the operator. NASA research pilot Herman Posada described flying the Ikhana (a civilian version of the MQ-9 Reaper) as "hours of boredom punctuated by a couple of moments of sheer terror during takeoff and landing."[5]

"There's a lot of stuff you're looking at while working the radios and checklists. It's a little too much for one (pilot). You need an extra set of eyes because sometimes you're drowning in information. Having other people say your speed is high or fast, or telling you to watch your sink rate is important."[6]

According to Posada, preparations for a typical Ikhana proficiency flight begin a few hours before dawn. The entire crew, accustomed to working from 7:30 a.m. to 4 p.m., adjusts scheduling to accommodate radio frequency allocations, which must be carefully managed to avoid conflict with other activities. "We show up at 4:00 a.m. to get NOTAMS [Notices to Airmen] and current weather, and then go to a crew briefing. Following preflight preparations in

---

3. Sharon D. Manning et al., "The Role of Human Causal Factors in U.S. Army Unmanned Aerial Vehicle Accidents," USAARL Report No. 2004-11, U.S. Army Aeromedical Research Laboratory, March 2004, *http://www.dtic.mil/cgi-bin/GetTRDoc?AD=ADA421592*, accessed July 27, 2011.

4. Ibid.

5. Peter W. Merlin, *Ikhana Unmanned Aircraft System Western States Fire Missions*, Monographs in Aerospace History #44 (Washington, DC: NASA SP-2009-4544, 2009).

6. Jay Levine, "No One on Board—Ikhana Pilots Fly Aircraft from the Ground," *X-tra*, 2008, *http://www.nasa.gov/centers/dryden/news/X-Press/stories/2008/07_08_pilots.html*, accessed July 27, 2011.

the GCS, the crew is ready to start engines around 5:45. The stress level goes up as soon as I add throttle for the takeoff roll."[7]

Posada must divide his concentration between flying the plane and running through "what if" scenarios in his mind. "What will I do if the engine fails during the takeoff roll? What if the engine fails immediately after takeoff?"[8]

Once airborne, he has to ensure that the airplane reaches proper altitude and remains clear of designated no-fly zones. Although a proficiency flight consists of just several touch-and-go landings for each pilot, maneuvering within the airport traffic pattern requires attention to airspeed, altitude, heading, communications, and aircraft configuration. As the aircraft approaches the runway, Posada checks the throttle setting and airspeed while intercepting the GPS Landing System localizer. He selects a visual reference point on the runway and maintains proper airspeed, transitioning to an idle power setting just before executing the landing flare. "Because there is no radar altimeter, determining the exact height over the runway is a bit of a challenge."[9]

In the back of his mind, Posada always has to consider potential go-around procedures in the event of pilot-induced oscillation or video dropout. Because of the high workload, each pilot usually performs no more than four touch-and-go practice landings. The final full-stop landing takes place about an hour after takeoff. "We try to fly these flights every week to stay current with the skills need to safely fly this type of aircraft."[10]

Automation of some or all vehicle systems is one way to reduce pilot workload, but it can also introduce several potential problems. The first of these is a loss of expertise, occurring whenever a machine replaces a human in performing a task. Reliance on a machine to perform a given function can degrade manual performance of the same task, should doing so become necessary as a result of mechanical malfunction. Another problem arises when operators become overconfident in the reliability of automated systems. This complacency may result in operators missing machine-based errors when they occur. Conversely, an unreliable automated system can lead to a lack of trust and system underutilization by operators. Finally, loss of adaptability may result from the automatic actions of the machine counteracting the operator's normal adaptive behavior.[11]

---

7. Personal communication from Herman posada, March 20, 2012.

8. Ibid.

9. Ibid.

10. Ibid.

11. Mark Rutley, "On Target—Human Factors: UAV's," Canadian Forces Environmental Medical Establishment, Jan. 2010, *http://www.airforce.forces.gc.ca/dfs-dsv/pub/nr-sp/index-eng. asp?id=10281*, accessed June 21, 2011.

Automation is clearly not a panacea for the challenges of flying remotely piloted aircraft. Rather than simply reducing overall workload, automation can redistribute existing work, introduce new and different tasks, and cause fundamental changes in the character of the pilot's workload. Additionally, automation forces new communications and coordination requirements on human operators.[12]

Low workload can pose as much of a problem as the opposing extreme. Some researchers have suggested that using a system that presents the operator with long periods of inactivity can produce vigilance-based stress. Tasks requiring constant attention, such as visually scanning electronic displays or running computer-assisted diagnostics, can induce a degraded attention span.

Vigilance can also be affected by haste, habit patterns, and complacency. In September 2000, an RQ-1L Predator crashed on the Nevada Test and Training Range after the pilot inadvertently activated a menu option that dumped the RAM memory in the airplane's primary control module. Investigators determined that as a result of a habit pattern developed over time and "a sense of being rushed," the pilot failed to verify the menu page and options to ensure that he was entering the correct command.[13]

Lance Prinzel, a senior research engineer and psychologist at NASA Langley, investigated the interaction between complacency and the use of automated systems, and he developed possible pilot workload strategies. He found that subjects with a higher motivation to successfully complete specific tasks were able to endure higher workloads.[14]

Some researchers have found that increased automation may negatively affect operator performance. Potential contributing factors include increased mental workload, loss of situational awareness, and skill degradation, as well as poor automation-to-user transition. Additionally, overconfidence in the capabilities of automated systems can lead to automation-induced complacency.[15]

---

12. Richard R. Sherry and Frank E. Ritter, "Dynamic Task Allocation: Issues for Implementing Adaptive Intelligent Automation," Technical Report No. ACS 2002-2, School of Information Sciences and Technology, Pennsylvania State University, July 2002, *http://acs.ist.psu.edu/misc/reports/sherryR02.pdf*, accessed June 16, 2011.

13. U.S. Air Force, "Executive Summary: Aircraft Accident Investigation, RQ-1L, S/N 96-3023, Indian Springs Air Force Auxiliary Field, NV, 14 September 2000," *http://usaf.aib.law.af.mil/ExecSum2000/RQ-1L_Nellis_14Sep00.pdf*, accessed June 21, 2011.

14. Manning et al., "The Role of Human Causal Factors in U.S. Army Unmanned Aerial Vehicle Accidents."

15. John J. Spravka, Deborah A. Moisio, and Mary G. Payton, "Unmanned Air Vehicles: A New Age in Human Factors Evaluations," in *Flight Test—Sharing Knowledge and Experience*, pp. 5-1 to 5-16, Meeting Proceedings RTO-MP-SCI-162, Paper 5A, Neuilly-sur-Seine, France, 2005.

MQ-1 Predator pilots occupy a typical ground control station. The Air Force operates RPA around the clock, requiring operators to work long shifts and odd hours. USAF

Physiological Factors: Fatigue and Stress

Weariness resulting from insufficient sleep, extended periods of mental or physical work, or prolonged periods of anxiety can affect RPA operators during long-duration missions. The human body has adapted to use sunrise and sunset to maintain consistent internal association of physiological functions with peak levels of activity timed to correspond with appropriate phases of the day-night cycle. Unfortunately, mission requirements may involve flying a vehicle throughout multiple days. Some remotely piloted vehicles are being designed to remain aloft for weeks or even months at a time. Operator fatigue levels may be compounded by continuous periods of work and fragmented sleep patterns. The resulting reduced reaction time and decreased vigilance can degrade performance, productivity, safety, and mission effectiveness.[16]

Both military and civil agencies operate RPA on long-duration missions. NASA scientific missions can last many hours while a vehicle is used to collect data. Remote sensing platforms supporting emergency response efforts are subject to deployment with short notice and at unpredictable times. The Air

16. Manning et al., "The Role of Human Causal Factors in U.S. Army Unmanned Aerial Vehicle Accidents."

Force and other Government agencies operate RPA around the clock in support of combat operations, requiring crews to work long shifts and odd hours. All of these factors may contribute to problems stemming from exhaustion.

According to a March 2008 study by the Naval Postgraduate School, RPA operators are more likely than other aircrews to suffer from chronic fatigue, emotional exhaustion, burnout, and impaired domestic relationships. A survey of 66 MQ-1 Predator crewmembers at Nellis Air Force Base, NV, found that nearly half of those surveyed suffered levels of fatigue during work hours that impaired job performance, and about 40 percent of those surveyed reported a moderate-to-high likelihood of falling asleep at their stations while operating a weapon-carrying remotely piloted aircraft. The study revealed that, "Pilots were found to have higher mental fatigue scores than sensor operators, suggesting a possible task-related contribution to their fatigue."[17]

Stress is also a factor in RPA operations. Although removed from physical hazards associated with flight, a remote pilot still has a sense of responsibility for the aircraft and the outcome of the mission. In 1969, NASA research pilot Milt Thompson found his first experience of flying a remotely piloted research vehicle surprisingly challenging. After the 3-minute flight, he was as physically and emotionally drained as he had been after piloting first flights in crewed research aircraft.[18]

As remotely piloted vehicles reached greater levels of sophistication, researchers at Dryden began monitoring the heart rates of their ground pilots. Data analysis included a comparison between a pilot's physical response to remote control and conventional flight. For example, when Einar Enevoldson flew the F-15 RPRV during its maiden flight, in 1973, he was required to execute 36 planned maneuvers as quickly as possible. The unpowered vehicle was dropped from the wing of a B-52 at 15,000 feet and flown as a glider while descending to 5,000 feet. Since the RPRV was recovered in midair, the pilot need not have worried about approach and landing—normally, one of the most stressful phases of flight. Enevoldson was known for his calm demeanor and a heart rate that typically remained below 80 beats per minute, even during hazardous crewed flights. Surprisingly, as the RPRV dropped away from the B-52, Enevoldson's heart rate jumped to 150 beats per minute before leveling off at 130 for most of the flight. One contributor to stress during this flight

---

17. Erik Holmes, "Study: UAV Crews Tired, Strained by Shortage," Air Force Times, April 18, 2008, *http://www.airforcetimes.com/news/2008/04/airforce_uav_fatigue_041808w/*, accessed June 16, 2011.

18. R. Dale Reed with Darlene Lister, *Wingless Flight: The Lifting Body Story* (Washington, DC: NASA SP-4220, 1997).

was the lack of physical cues, or what Enevoldson described as "the essential verifying, comforting sensations of flight."[19]

Ikhana pilot Mark Pestana noted that this absence of physical sensations is a source of stress even for experienced RPA pilots during routine operations such as proficiency flights. The primary objective of such flights is to allow pilots to practice takeoff and landing—typically, the most difficult aspects of manually flying the MQ-9 because the airplane's autopilot has no automatic takeoff or landing capability. A human pilot must use the traditional control stick, rudder pedals, and engine throttle for taxi, takeoff, approach, and landing. Flying the MQ-9 from a GCS, the pilot has only a limited camera view of the environment and none of the sensory feedback normally associated with flying an airplane (i.e., vestibular cues, vibrations, engine sounds, and peripheral vision cues that aid in sensing speed, descent/climb rate, and height above ground). "Typically in manned aircraft, when a pilot makes adjustments to engine power and attitude, feedback is provided via a combination of physical sensations. Lacking these cues, the RPA pilot can become task-saturated while dividing attention between executing tasks and assessing aircraft response while simultaneously maintaining situational awareness and verifying the airplane's actual response to control inputs, and applying corrective actions…all in a continuous loop."[20]

During final approach, the pilot must make flight-control and power adjustments to maintain glideslope, descent rate, and approach speed while rapidly scanning various indicators (airspeed, engine torque, glideslope, and aim-point). This process of continuously making control inputs and verifying aircraft response and status can create a stressful, task-saturated environment in the cockpit. In the final moments before landing, the pilot makes precise adjustments to maintain proper approach speed and angle and begins to raise the pitch angle for landing flare and reduce engine power for touchdown. During Ikhana flights, the pilot must also repeatedly update a "lost link" start point to ensure that the airplane will fly a preplanned route to a safe holding pattern away from the airfield in the event of uplink failure. "In this rapid, multi-task environment, these few moments of diverted attention and additional tasks seem like an eternity when the constant task of maintaining controlled flight is repeatedly interrupted."[21]

---

19. Terrence W. Rezek, "Unmanned Vehicle Systems Experiences at the Dryden Flight Research Facility," NASA TM-84913 (June 1983).

20. Personal communication from Mark Pestana, March 20, 2012.

21. Ibid.

The Ikhana is not equipped with precision altimetry equipment, and it relies on a barometric altimeter with inherent limitations in latency and accuracy. According to Pestana, the pilot performs the final flare, power reduction, and landing using visual reference to the camera view (what he calls the TLAR, or the "That Looks About Right" method). "These unconventional procedures and tasks, each requiring concentrated visual attention to multiple sources of information in non-standard formats, can saturate the pilot to the point that, after only three or four trips around the traffic pattern, he is ready to stop. Even a 0.8-hour proficiency flight results in adrenaline rush and fatigue."[22]

Environmental Factors: Situational Awareness

Lack of conscious, as well as subconscious, cues remains a challenge for RPA operators because it affects situational awareness. The entire experience of flying remotely is very different compared with conventional flying because the RPA pilot lacks such physical cues as visibility, motion, sound, feel, and even smell. When Mark Pestana joined the Ikhana program, he had over 4,000 flight hours in numerous aircraft, but none in RPA. He found the lack of vestibular cues and peripheral vision hindered his situational awareness and eliminated his ability to experience such sensations as motion and sink rate.

"It was like I had lost four of my five senses," he said of his experience in the ground cockpit. "Your vision is limited because there is only a single camera for forward visibility, you can't hear the engine, feel the aircraft's motion or acceleration, or smell a fuel leak or an electrical fire."[23]

Pestana discovered that this aspect of RPA operations was particularly challenging during landing because the pilot has no direct view of the runway until turning onto final approach. Instead of maintaining situational awareness while flying crosswind, downwind, and base leg by looking out through a wide field of windows to maintain a reference position to the runway, the RPA pilot must pay attention to a map screen with a moving icon representing the airplane. "This screen is mounted well above the typical instrument viewing angles of a manned aircraft's control console, drawing the pilot's attention away from the narrow-field camera view and primary flight instrument displays."[24]

Degraded situational awareness as a cause of aircraft accidents results from the failure to correctly perceive information, failure to integrate or comprehend

---

22. Ibid.

23. Merlin, *Ikhana Unmanned Aircraft System Western States Fire Missions.*

24. Personal communication from Mark Pestana, March 20, 2012.

information, or failure to project future actions or system states. RPA operators are subject to impaired situational awareness as a result of several factors. The first is the lack of a 360-degree view from the cockpit that ordinarily provides the pilot with the ability to inspect the aircraft for damage or icing, watch for traffic, and assess weather conditions or terrain hazards.

Some vehicles are equipped with one or two cameras that only provide a narrow field of view (sometimes called a "soda straw" view) of the environment through which the aircraft is flying. Images from these cameras are displayed on a monitor in the pilot's station. Vehicles such as the RQ-4 Global Hawk provide no external view whatsoever. The operator has only an onscreen map with an icon representing the aircraft's location. Consequently, the pilot lacks visual information such as peripheral vision of surrounding airspace and "ground rush" during landing approach. This situation is mitigated to some extent on aircraft such as the Predator that are equipped with a rotating "Sky Ball" camera/sensor turret. As noted earlier, the pilot lacks such physical cues as kinesthetic and vestibular input that ordinarily provide a "seat of the pants" feel of g-forces as well as auditory input including engine and wind noise.[25]

To compound an already difficult situation for the pilot, the physical distance between pilot and aircraft leads to delays in uplink/downlink signal transmission. Delays of 1 second or more can introduce temporal and spatial uncertainties. With delays between pilot input and aircraft feedback, errors in the operator's commands and responses can result in total loss of control of the vehicle.[26]

Data link bandwidth limitations can reduce the RPA operator's temporal resolution, spatial resolution, and color discrimination capabilities. Image update lag time due to bandwidth limits also inhibits situational awareness. All of these factors impose extra mental processing requirements and affect the operator's ability to build and maintain an adequate mental model of the situation.

Lack of situational awareness with regard to weather conditions led to the loss of an RQ-1L Predator in Southwest Asia on September 17, 2002. While an operator at Nellis Air Force Base was performing a reconnaissance mission in support of Operation Enduring Freedom, the aircraft entered an area of meteorological convective activity. The satellite link was interrupted for approximately 20 seconds, after which the aircraft was unresponsive to commands. The primary cause of the mishap was attributed to the pilot flying

---

25. Rutley, "On Target—Human Factors: UAV's."

26. Manning et al., "The Role of Human Causal Factors in U.S. Army Unmanned Aerial Vehicle Accidents."

the RPA unintentionally into the type of cloud where the likelihood of severe turbulence was high. A pilot on board a light aircraft would typically avoid such circumstances, but the Predator pilot was located on the other side of the world and had no ability to make direct observations of local weather conditions.[27]

Too much focus on mission objectives can also reduce situational awareness. On October 3, 2009, two crews were involved in the loss of an MQ-1B Predator supporting combat operations in Afghanistan. After the Predator departed on a reconnaissance mission, Crew 1 received orders to provide close air support for U.S. and Afghan forces under attack in two locations. While en route, Crew 2 assumed control of the Predator from Crew 1. By this time, several U.S. troops had been killed in the attacks. Given the circumstances and the immediate need for close air support, both RPA crews were consumed with a high degree of urgency. Neither crew perceived any environmental threat. Both crews failed to appreciate the need for a significant increase in altitude required to safely overfly the mountainous terrain between the aircraft and the target. Investigators subsequently determined that the Crew 2 pilot's channelized attention on reaching the target and both crews' inattention to high mountains in the immediate vicinity led to controlled flight into terrain.[28]

Channelized attention can affect routine operations as well. On August 31, 2010, an MQ-9A Reaper crashed 2 miles north of Gray Butte Flight Test Station, Palmdale, CA, during a training mission. While the pilot was attempting to perform a low-speed, high-angle-of-attack maneuver, his head-up display showed a calculated stall speed of 69 knots indicated airspeed. During the attempted maneuver, the Reaper's speed dropped as low as 61 KIAS, below stall speed, for 8 seconds, resulting in the aircraft stalling and rolling into a left-hand spin. Although the pilot initiated emergency procedures for stall and spin recovery, the aircraft's rate of descent reached 5,000 feet per minute. Before the pilot could recover control, the Reaper crashed. Investigators found convincing evidence of pilot error, indicating that the pilot's attention was focused on setting up the angle of attack for the slow-flight maneuver. Additionally, the General Atomics flight-training program for the Reaper did not adequately

27. U.S. Air Force, "Executive Summary: Aircraft Accident Investigation, RQ-1L 'Predator' S/N 96-3024, Nellis Air Force Base, NV, 17 September 2002," *http://usaf.aib.law.af.mil/RQ-1L_SouthwestAsia_17Sep02.pdf*, accessed June 10, 2011.

28. U.S. Air Force, "Executive Summary: Aircraft Investigation, MQ-1B, T/N 06-3175, March Joint Air Reserve Base, 3 October 2009," *http://usaf.aib.law.af.mil/ExecSum2010/MQ1B_3%20 Oct%2009.pdf*, accessed June 10, 2011.

With vehicles like the MQ-1, proper crew composition, selection, and training are essential to safe and efficient flight operations. USAF

prepare the pilot for the risks associated with performing slow flight at angles of attack beyond those previously demonstrated during flight test.[29]

Several methods may help overcome situational awareness deficiencies. Varying degrees of automation can compensate for the fact that the operator is not colocated with the aircraft. A multimodal control interface can provide the operator with enhanced situational awareness through the use of alarms, warning lights, and touch-based ("stick shaker") systems. Anything that reduces the operator's mental workload will free up additional mental resources, improving task efficiency and flight safety.[30]

29. U.S. Air Force, "Executive Summary: Aircraft Accident Investigation, MQ-9A, T/N 09-4056, Edwards Air Force Base, CA, 31 August 2010," *http://usaf.aib.law.af.mil/ExecSum2010/ MQ9A_31%20Aug%2010.pdf*, accessed June 10, 2011.

30. Rutley, "On Target—Human Factors: UAV's."

Staffing Factors: Training and Crew Coordination

Good training and teamwork are critical to the safe operation of remotely piloted vehicles. Elements of an effective training program include the use of experienced instructors, well-defined standards, and an effective evaluation process. Instructors should be sure to emphasize the value of effective crew coordination. Inadequate training, failure to follow established procedure, and lack of crew coordination are common factors in RPA mishaps.[31]

In one example, on January 22, 2002, an RQ-1L Predator departed controlled flight and crashed shortly after takeoff from a forward operating location in Southwest Asia. The mission plan included a control handoff of the Predator between two ground control stations using a lost-link recovery orbit point. While attempting to execute the handover, one crew failed to perform checklist steps in the proper order, accidentally shutting off the Predator's engine and stability augmentation system. The crew was unable to fully recover from the checklist error, and the vehicle crashed. The cause was attributed to training deficiencies and lack of crew coordination.[32]

Another crew at a different operating location suffered a similar mishap on November 24, 2004. While being operated by contractor personnel, an MQ-1L crashed during recovery following an in-flight navigation malfunction. After aligning the aircraft for final approach, the pilot realized that the touchdown point would be short of the runway. He attempted to raise the nose but discovered that he could not change the aircraft's nose-down attitude, and the Predator crashed 300 feet short of the runway. Investigators determined that the pilot had not properly executed his landing checklist, failing to disengage the autopilot's airspeed hold mode. As a result, the aircraft maintained its programmed airspeed and flew into the ground. Training, flight discipline, and supervision issues contributed to pilot error in this case.[33]

In another incident involving procedural errors, an MQ-1B Predator departed controlled flight and crashed northeast of Kandahar Air Base, Afghanistan, on December 9, 2010. Descending toward the airfield, the pilot, intending to press the "Landing Configuration Switch" (this is a badly misleading name for the

---

31. Manning et al., "The Role of Human Causal Factors in U.S. Army Unmanned Aerial Vehicle Accidents."

32. U.S. Air Force, "Executive Summary: Aircraft Accident Investigation, RQ-1I Predator, S/N 97-3031, undisclosed forward operating location in support of Operation Enduring Freedom, 22 January 2002," *http://usaf.aib.law.af.mil/RQ-1L_FOL_22Jan02.pdf*, accessed June 11, 2011.

33. U.S. Air Force, "Executive Summary: Aircraft Accident Investigation, RQ-1I Predator, S/N 97-3031, undisclosed forward operating location in support of Operation Enduring Freedom, 22 January 2002," *http://usaf.aib.law.af.mil/RQ-1L_FOL_22Jan02.pdf*, accessed June 11, 2011.

Autopilot Disconnect), inadvertently disabled the Stability Augmentation System (SAS) by pressing the SAS switch, located on the top right side of the control stick. The vehicle pitched down, and the pilot responded with a command to raise the nose. Excessive pitch-up led to a stall, after which the aircraft dove toward the ground. Contrary to procedures established in the flight manual, the pilot inadvertently turned on the SAS, which should have been done only after returning the aircraft to straight and level flight. When the SAS was activated outside its recommended range, internal software logic interpreted the aircraft's current attitude (and its associated control-surface deflection) as the intended parameters and set pitch trim accordingly. The pilot attempted to level the aircraft by pulling back on the stick but never attempted to change the pitch-trim setting. Simultaneous application of these inputs would have allowed the Predator to achieve level flight. Although the mishap sequence began with the inadvertent use of the SAS switch, investigators determined that the primary cause was the pilot's failure to apply the proper stall recovery procedures.[34]

Proper composition, selection, and training of RPA crews are essential to safe and efficient flight operations. Typically, flightcrews for remotely piloted aircraft include at least one pilot (often with a backup or copilot) with responsibility for control of the aircraft and one or more systems operators to monitor and control payloads and sensors. Crewmembers should always make an effort to follow proper procedures, but RPA aircrew training must also emphasize crew coordination. The principles, attitudes, procedures, and techniques that transform individuals into an effective crew are a vital part of any RPA training program. Coordinated efforts by the entire crew ensure safety and effective performance.[35]

The crew staffing a typical RPA ground station may consist of operators/pilots, systems monitors, and mission coordinators or test conductors. Crew coordination and teamwork includes interaction (positive communication) between crewmembers and actions (sequence and timing) necessary to perform tasks efficiently, effectively, and safely. Communication is positive when the sender directs, announces, requests, or offers, the receiver acknowledges, and

34. U.S. Air Force, "Executive Summary: Aircraft Accident Investigation, MQ-1B Predator, S/N 05-3137, near Kandahar Airfield, Afghanistan, 9 December 2010," *http://usaf.aib.law.af.mil/ExecSum2011/MQ-1B_Kandahar_9%20Dec%2010.pdf*, accessed June 11, 2011.

35. Jason S. McCarley and Christopher D. Wickens, "Human Factors Concerns in UAV Flight," Technical Report AHFD-05-05/FAA-05-01, Institute of Aviation, Aviation Human Factors Division, University of Illinois at Urbana-Champaign, IL (2005).

the sender confirms. Positive communication is quickly and clearly understood and permits timely actions.[36]

Design Factors: Human-Machine Interface

The human-machine interface as it relates to remotely piloted aircraft involves physically fitting the operator to the control station. Relevant design factors include displays to provide situational awareness, control layout, input devices (buttons, switches, etc.), ambient lighting, and seating configuration. Since RPA operation entails large amounts of data, control station designers must take into account human limitations for processing information. Displays need to present data in such a manner as to allow for efficient interpretation by the operator.[37]

Though human error has been blamed for numerous RPA mishaps, a closer examination reveals that many of these errors result directly from design shortfalls in the human-machine interface. Lack of privacy is one of the most serious problems. NASA MQ-9 pilot Mark Pestana noted that unlike a normal cockpit, a GCS is not a "private office in the sky."[38] People come and go, opening and closing doors and holding casual conversations. Ringing telephones, whispered remarks, and other disturbances can interrupt critical operations—such as approach and landing maneuvers—that demand silence and concentration.

Any interface design must account for operator attention issues, such as diversion from a task during critical moments or becoming channelized. The latter, also known as cognitive tunneling, occurs when an operator's attention is focused on one task to the detriment of all others. Control characteristics of the RPA or ground control station should be sufficiently forgiving to prevent catastrophic failure in the event that the pilot misses making a critical control input on time. Additionally, when the operator returns attention to control displays, the state of the aircraft should be immediately recognizable so that proper action may be taken. The human-machine interface should be designed

---

36. *Unmanned Aerial Vehicle Aircrew Training Manual* (TC 34-212) (Washington, DC: Headquarters, Department of the Army, August 1997), *http://www.fas.org/irp/doddir/army/34-212.pdf*, accessed June 10, 2011.

37. Manning et al., "The Role of Human Causal Factors in U.S. Army Unmanned Aerial Vehicle Accidents."

38. Mark E. Pestana, "Flying NASA Unmanned Aircraft: A Pilot's Perspective," AIAA Paper 2011-1490, presented at Infotech@Aerospace 2011: Unleashing Unmanned Systems, St. Louis, MO, March 30, 2011, p. 4.

to be as simple as possible to avoid overloading the operator with information or tasks. Detailed information must be available to the operator at various points during performance of tasks, but overall situational awareness must be maintained at all times.[39]

Since pilot workload while flying a remotely piloted aircraft is primarily mental in nature, maintaining situational awareness through the human-machine interface is crucial. With the exception of most cockpit environmental factors (life support, vibration, noise, temperature), human-system-integration issues for remotely piloted aircraft are essentially the same as those for crewed aircraft. Cockpit displays for both must be readable and easily interpreted. Cockpit ergonomics must include good body clearance, easily reachable controls, and appropriate display visibility. Controls must be appropriately located and labeled, they must be arranged for ease of use, anthropometric accommodation, and discriminability between colocated controls. Communications should be clear and intelligible and follow standard aviation phraseology and terminology. Information displays should be optimized for situational awareness. Finally, the type, complexity, duration, difficulty, and resource demands of any given task or combination of tasks should not result in workload exceeding the operator's capabilities.[40]

The use of system automation is increasing for both crewed and uncrewed aircraft. Designers must ensure that ground control workstations offer the appropriate degree of automation and level of control to the RPA operator. Levels of control range from total to supervisory.

In a system subject to entirely manual control (human-in-the-loop), the human operator specifies the functions to be accomplished and the machine carries out the tasks as directed. Such operations typically involve ground stations equipped with traditional aircraft controls such as stick and throttle. The RPA pilot may also rely on aircraft-mounted video cameras to provide visual information during takeoff, flight, and landing. Additional visual displays provide the pilot and/or systems monitors with aircraft subsystem health, navigation, and sensor information. The MQ-1 and MQ-9 are good examples of systems that feature manual control.[41]

In supervisory control (human-on-the-loop), the human operator specifies the goals but the machine carries out tasks and functions autonomously. The operator enters macro level commands via mouse and/or keyboard to direct the aircraft through a series of flight operations. Supervisory control tasking may include clicking screen icons or making keyboard entries to initiate engine

---

39. Rutley, "On Target—Human Factors: UAV's."

40. Spravka, Moisio, and Payton, "Unmanned Air Vehicles: A New Age in Human Factors Evaluations."

41. Ibid.

start and takeoff, navigate between waypoints, change airspeed, or select an entirely new mission. The Global Hawk system is an example of one that features supervisory control. The pilot provides high-level command input but does not have stick-and-throttle control.

In 2003, John M. Reising of the Air Force Research Laboratory at Wright-Patterson Air Force Base, OH, proposed developing RPA systems exhibiting a "cooperative functioning control"[42] architecture in which human and machine interact at all levels to accomplish goals, functions, and tasks. Although the human operator would maintain ultimate control over the entire system, both human and machine would dynamically share all levels of control authority throughout the mission. Aircraft systems operating at a true cooperative functioning control level have not yet been developed.

For some RPA systems, an external pilot (the "outside pilot") is responsible for takeoff and landing. This individual is physically located outside the GCS and maintains visual contact with the aircraft during takeoff and landing operations. After takeoff, control is handed off to a pilot inside the GCS. At the end of the flight, the inside pilot transfers control to the outside pilot prior to landing.

External pilots typically operate controls resembling those used with radio-controlled model aircraft, using joysticks on a handheld or stationary console to manipulate the flight-control surfaces of the RPA. External pilots require skills similar to those of the model aircraft enthusiast, and experience the same orientation issues that require superior three-dimensional spatial reasoning skills. For example, when the controller and the aircraft are facing the same direction, a rightward joystick input correlates to rightward motion relative to the controller. However, when the controller and aircraft face in opposite directions, the same input correlates to a leftward motion.[43]

Flying the X-36

The importance of operator interface was highlighted during the X-36 research program at Dryden in the 1990s. The X-36 was a 28-percent-scale representation of a generic, advanced-tailless, agile, stealthy fighter aircraft configuration.

---

42. Ibid.

43. Michael Nas, "The Changing Face of the Interface: An Overview of UAS Control Issues and Controller Certification," Unmanned Aircraft Technology Applications Research Working Group 27 (UATAR WG-27), Feb. 20, 2008, *http://www.uatar.com/UAS%20Control%20Issues%20-%20UATAR%20(2).pdf*, accessed Sept. 27, 2011.

The X-36 was a remotely piloted, subscale, tailless aircraft technology demonstrator. NASA

It was approximately 18 feet long with a wingspan of just over 10 feet. A single Williams International F112 turbofan engine provided about 700 pounds of thrust. Fully fueled, the X-36 weighed 1,250 pounds. Due to the advanced configuration, researchers considered this a high-risk program. Safety risks were minimized by eliminating the need for an onboard pilot, but programmatic risks were relatively high due to the increased probability of losing the aircraft in a mishap.

These risks were mitigated by using a pilot-in-the-loop approach to eliminate the need for expensive and complex autonomous flight-control systems and hazards associated with their inability to react properly to unknown or unforeseen phenomena once in flight. Boeing project pilot Laurence Walker was a strong advocate for the advantages of using a full-size ground cockpit. When designing a control station for a subscale remotely piloted vehicle there is a natural tendency to reduce the cockpit control and display suite, but in a 1997 analysis, Walker suggested that the best practice is just the opposite. As previously noted, in any ground-based cockpit, the pilot will have fewer natural sensory cues such as peripheral vision, sound, and motion. Recreating motion cues for the X-36 was impractical, but audio, visual, and head-up display (HUD) cues were recreated in order to improve overall situational awareness comparable to that in a full-size aircraft.[44]

---

44. Laurence A. Walker, "Flight Testing the X-36—The Test Pilot's Perspective," NASA CR 198058 (1997).

The X-36 ground control station included a full-size control stick, rudder pedals, throttle, and a full complement of modern fighter-style switches. A video camera mounted on the airplane provided situational awareness data to the ground pilot, along with a moving-map display. NASA

The X-36 ground control station included a full-size control stick, rudder pedals, throttle, and a full complement of modern fighter-style switches. Situational awareness data were provided to the ground pilot through a video camera mounted in the nose of the vehicle, a standard fighter-type HUD, and a moving-map representation of the vehicle's position within the range. Visual data were displayed on two 20-inch monitors. The forward-looking monitor provided downlinked video from the canopy-mounted camera as well as HUD overlay with embedded flight-test features. The second monitor showed a "God's-eye-view" horizontal situation indicator, engine and fuel displays, control-surface-deflection indicators, yaw rate, and a host of warnings, cautions, and advisories. An audio alarm alerted the pilot to any new warnings or cautions. A redundant monitor shared by the test director and GCS engineer served as a backup, should either of the pilot's monitors fail.[45]

To improve the pilot's ability to accurately set engine power and to further improve situational awareness, the X-36 vehicle was equipped with a microphone in what would have been the cockpit area of a conventional aircraft. Downlinked audio of engine noise from this microphone proved to be a highly valuable cue

---

45. "Fact Sheets—X-36 Tailless Fighter Agility Research Aircraft," NASA Web site, 1999, *http://www.nasa.gov/centers/dryden/news/FactSheets/FS-065-DFRC.html*, accessed Sept. 15, 2011.

and alerted the team, more than once, to problems such as engine screech at high power settings and engine stalls before they became serious.

The X-36 had a very high roll rate and a mild spiral divergence. Because of its size, it was also highly susceptible to wind gusts. As a result, the pilot had to spend a great deal of time watching the HUD, the sole source of attitude cues. Without kinesthetic cues to signal a deviation, anything taking the pilot's focus away from the HUD (such as shuffling test cards on a kneeboard) was a dangerous distraction. Based on flight-test experience, the X-36 team solved this problem by designing a tray to hold test cards at the lower edge of the HUD monitor for easy viewing. The cards were arranged like a hand of playing cards so that the pilot could pull the top sheet off without looking away from the HUD.

In order to allow the pilot to keep his attention focused on his displays, the crew station at the center of the GCS was isolated beneath a tent-like shelter. The rear section, behind the pilot, was left uncovered so that a flight-test engineer could monitor aircraft systems and assist as necessary. To further reduce distractions, a dedicated communications loop was established to include only the pilot, test director, and outside radio communications. Technology engineers were left out of the loop so they could be free to discuss test results and anomalies without disturbing the pilot.[46]

When the first flight was flown on May 17, 1997, the X-36 team found no obvious deficiencies. The second flight, however, uncovered a significant problem. When the X-36 was about 10 miles from the GCS and at 12,000 feet altitude, the video and downlink signals suddenly became very weak with static and video noise interference. The pilot suddenly found himself staring at a frozen display with a big "X" indicating that the X-36 had gone into lost-link autonomous operation. The test team immediately began recovery procedures to regain control, and the pilot managed to reestablish the link momentarily on several occasions only to lose it seconds later. Each intermittent glimpse of the data showed the vehicle in a steeper angle of bank, well beyond any maneuver that had yet been flown. Eventually, the ground pilot regained control and executed an uneventful landing. Throughout the ensuing test program, flying the X-36 proved a very work-intensive experience for the ground pilot, but the ground cockpit design proved extremely helpful.

Walker later reported, "I was very glad that we had invested the extra effort to provide good situational awareness and minimize pilot distraction."[47]

---

46. Walker, "Flight Testing the X-36—The Test Pilot's Perspective."
47. Ibid.

AeroVironment developed a pilot's station similar to those typically used to fly remote-controlled models. Here, pilot Wyatt Sadler uses the "flight station in a box" to fly the Pathfinder. NASA

Controls and Displays

Ground cockpit designs vary significantly from one vehicle to the next. Factors determining the differences include the heritage of the aircraft's design as well as the experience and resources of the operating organization. Performance characteristics of the X-36, such as high landing speed and long rollout distances, led designers to choose a pilot-in-the-seat approach in which one person flies the airplane from a virtual cockpit.

The Helios test team used a completely different philosophy. With extensive experience flying remote-controlled models, AeroVironment's team developed a pilot's station more like what one would expect from someone accustomed to flying the airplane from the outside. Although the control suite provided ample displays for flight control, navigation, and systems monitoring, the stick looked similar to that used for hobby radio-controlled models. Additionally, downlinked video did not provide a "pilot's eye view," as from a conventional cockpit. Instead, because it was primarily used for monitoring the vehicle's structure, motors, and propellers, it looked inboard across the airframe from the left wingtip. As a result, two crews were needed to fly the airplane. One

NASA pilots Mark Pestana (left) and Herman Posada sit at their stations in the Ikhana ground control station at Dryden. The General Atomics GCS includes many features found in a conventional airplane cockpit. NASA

crew performed takeoff while directly observing the Helios from the edge of the runway, physically observing the airplane's orientation and proximity to the ground, noting any problems and observing any potential traffic conflict. As the vehicle approached the visual limits of the first crew, control was transferred to a second crew in a ground control station. Upon return to the airfield, control of the vehicle was returned to the first crew for approach and landing. AeroVironment successfully demonstrated this system during both day and night operations.[48]

Because of the airplane's design heritage, the human-machine interface for NASA's Ikhana is based on the Predator/Reaper ground control station. The GCS cockpit features pilot and payload operator stations, the latter of which also serves as a copilot station with redundant controls. Multiple CRT screens provide a head-up display with options for a military-style data presentation,

---

48. John H. Del Frate and Gary B. Cosentino, "Recent Flight Test Experience with Uninhabited Aerial Vehicles at the NASA Dryden Flight Research Center," NASA TM-1998-206546 (April 1998).

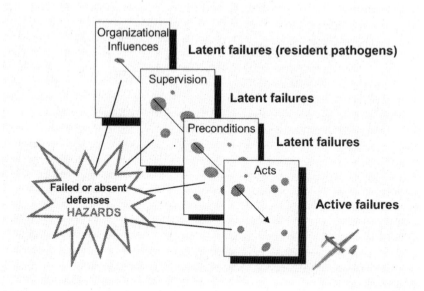

James Reason's "Swiss cheese" model demonstrates how latent failures contribute to mishaps. NASA

one more like what might be found in a civilian light aircraft (referred to by some pilots as the "Cessna display"). Only one pilot at a time can fly the aircraft, although the copilot assists during the demanding takeoff and landing phases and can issue commands to configure the various aircraft systems as necessary. Unlike in a conventional airplane, control inputs are made with a keyboard and joystick. When one pilot needs to take a break, another can take over so that there is always a fully alert crewmember at the controls.[49]

The flight controls are markedly different from those of a conventional aircraft. The pilot's station includes a set of rudder/brake pedals, engine throttle, propeller controls, and a control stick grip, but the similarities end there. Instead of using control stick deflection to set pitch or bank at a steady rate of change, the Ikhana stick deflection sets a specific fixed pitch or bank angle. Instead of a standard control panel, the pilot uses two systems display screens to access more than 60 pages of data.[50]

"Instead of physical switches—toggle switches or dials—you're using a keyboard and trackball and pulling down menus like you would on your personal computer to activate systems," NASA pilot Mark Pestana explained.

---

49. Merlin, *Ikhana Unmanned Aircraft System Western States Fire Missions.*

50. Jay Levine, "No One on Board—Ikhana Pilots Fly Aircraft from the Ground."

"Understanding where all of these system controls are located, and finding the right screen display to access the controls, is challenging."[51]

RPA operators, whether in civilian or military applications, have widely varying levels of flying experience. The Army typically employs enlisted personnel with some basic flight training, while the Air Force has traditionally used commissioned pilots pulled from fighter, bomber, and transport units. With increasing use of remotely piloted platforms resulting in increased demands for operators, designers are working to develop improved human-machine interfaces as well as techniques for a single operator to control multiple vehicles simultaneously. Future options for ground control display systems range from conventional 2D video screens to flat perspective-view visualizations to three-dimensional immersive displays and multisensory interface technology. An advanced interface incorporating improved human-factors engineering should increase the operator's situational awareness, improve workload management, and improve overall system performance.[52]

---

51. Ibid.

52. J.R. Wilson, "UAVs and the Human Factor," *Aerospace America Online*, July 2002, *http://www.aiaa. org/aerospace/Article.cfm?issuetocid=233&ArchiveIssueID=28,* accessed June 16, 2011.

A U.S. Customs and Border Patrol Agency MQ-9 patrols the U.S.-Mexico border in southern Arizona. CBP

CHAPTER 9

Analysis of Human Factors in a Predator-B Crash

As remotely piloted aircraft become more commonplace, many countries, including the United States, are seeking ways to integrate them into the national airspace. Early efforts in this direction include the use of remotely piloted platforms for customs and border patrol missions along the U.S.-Mexico border. The April 25, 2006, crash of a U.S. Customs and Border Protection Agency (CBP) MQ-9 Predator-B near Nogales, AZ, fueled concerns over the safety of remotely piloted aircraft and triggered a painstaking investigation led by the National Transportation Safety Board (NTSB). Although the direct causes were easily identified, a subsequent study by several MIT researchers used the Human Factors Analysis and Classification System (HFACS) to identify numerous latent errors that contributed to the accident but were not addressed during the NTSB investigation.[1]

"Switchology" Error

At the time of the mishap, the MQ-9 was owned by the CBP and operated under contract with General Atomics Aeronautical Systems, Inc., manufacturer of the aircraft and ground control station. This arrangement allowed the CBP to expedite the employment of remotely piloted surveillance aircraft to improve security along the southern border of the United States.[2]

The turboprop-powered MQ-9 was designed as a long-endurance, high-altitude platform with a 66-foot wingspan, maximum gross weight of 10,000 pounds, and a top speed in excess of 220 knots. It is equipped with

---

1. Chris W. Johnson and Christine Shea, "The Hidden Human Factors in Unmanned Aerial Vehicles," in *Proceedings of the 26th International Conference on Systems Safety* (Unionville, VA: International Systems Safety Society, 2008).

2. Ibid.

redundant, fault-tolerant avionics and is typically flown by two pilot payload operators. Normally, a certified pilot occupies the left seat (PPO-1) while a U.S. Border Patrol agent operates the surveillance camera from the right seat console (PPO-2). The side-by-side consoles have nearly identical sets of controls (flaps, throttle, airspeed, condition lever, monitors), but depending on which console is being used to fly the aircraft, these controls may be set to alternate modes.

When PPO-1 is used as the primary aircraft control interface, movement of a device known as the "condition lever" to the forward position opens the fuel valve to feed the engine. The pilot shuts down the engine by moving that same lever back to the middle position, closing the fuel valve. Moving the lever full aft feathers the propeller. When PPO-1 is used to fly the aircraft, PPO-2 serves as a camera control console and its condition lever controls the lens iris setting. This lever is usually set in the middle position to lock the camera's iris setting.[3]

The CBP flight operation was based at Libby Army Airfield, near Sierra Vista, AZ. Takeoff was scheduled for the early evening of April 24 but was slightly delayed because of the inability to establish a communication link between the aircraft and PPO-1 during initial power-up. An avionics technician powered the aircraft down, downloaded system status, and then recycled the power. Still unable to establish a communications link with PPO-1, he contacted his supervisor at the manufacturer's California facility for assistance. The supervisor recommended switching the main processor cards between PPO-1 and PPO-2. The technician did this, powered up the system, and successfully established an uplink with both PPO-1 and PPO-2. The MQ-9 took off without incident.

A relief pilot took the controls a few hours later, after the MQ-9 had climbed to altitude and was operating within temporary flight-restriction airspace used for RPA operations. The first pilot resumed control of the aircraft 2 hours later, and the relief pilot replaced him 4 hours after that, at around 3 a.m. Shortly after taking control, the relief pilot watched his monitor screen go blank. When the telemetry data reappeared, it was locked.

The avionics technician, seated at his multifunction workstation, noticed that his monitor had also locked up. He told the pilot to switch control to PPO-2 and take over flying from that console. He then opened a program to record information on the status of PPO-1.[4]

---

3. "NTSB Accident Brief: General Atomic Predator-B, April 25, 2006, Nogales, AZ (Docket CHI06MA121)," *http://www.ntsb.gov/NTSB/brief.asp?ev_id=20060509X00531&key=1*, accessed June 10, 2011.

4. Ibid.

CBP agents operated the MQ-9 from a standard GCS, with identical consoles for pilot and sensor operator. CBP

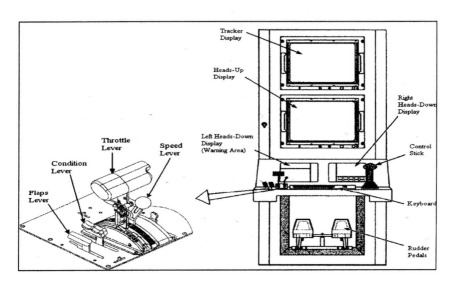

When PPO-1 is used as the primary aircraft control interface, movement of the "condition lever" to the forward position opens and closes the fuel valve. Moving the lever full aft closes the fuel valve and feathers the propeller. NTSB

The pilot informed the Border Patrol agent seated at PPO-2 that he needed to switch stations, and he verified that the ignition was "hot" on PPO-2 and the stability augmentation system was on. He then used his cell phone to call another pilot (who had been his instructor) to discuss the situation. The avionics technician again advised the pilot to switch to PPO-2. By this point, the Border Patrol agent had departed the ground control station.

The MQ-9 flight checklist required that qualified pilots be seated at both the PPO-1 and PPO-2 positions prior to switching control of the aircraft from one to the other. This did not happen simply because there was only one pilot in the trailer at the time. The pilot later stated that he did not consult the checklist during switchover.

According to the checklist, before switching control from one station to another, the primary pilot is supposed to match control positions on both consoles. In this case, the pilot failed to do this, he later told investigators, because he was "in a hurry."[5] As a result, the situation lever, which had been used to lock the camera iris, was now in the fuel shutoff position. In aviation, this is known as a "switchology" error. The term refers to techniques used to ensure the proper settings of switches on control panels.

When the pilot switched control to PPO-2, the condition lever was still centered. In camera-control mode, this locked the iris, but in flight-control mode, it cut off fuel flow to the engine. As the engine sputtered to a halt, the MQ-9 began to descend. The pilot noticed the change in altitude but failed to discern the cause. Also unaware of the situation, the avionics technician advised him to shut down the ground-data terminal so that the MQ-9 would begin its programmed lost-link procedure. Normally, this would cause the MQ-9 to autonomously climb to 15,000 feet and fly a predetermined course until uplink could be reestablished. Without engine power, however, the aircraft descended until it was below the range of line-of-sight communications and contact was lost. The MQ-9 crashed in a sparsely populated residential area, but there were no injuries or damage to private property.[6]

NTSB Investigation

This was the first time the National Transportation Safety Board investigated an accident involving an uncrewed aircraft, principally because it took place in national rather than military airspace. On October 16, 2006, NTSB officials

---

5. Ibid.
6. Ibid.

The MQ-9 crashed in a sparsely populated residential area, but there were no injuries or damage to private property. Mae Duggin

announced that operator error was the primary cause of the mishap near Nogales. A report summarizing the board's conclusions noted that the probable cause of the accident was the pilot's failure to use checklist procedures when switching control from a console that became inoperable due to a control console lockup, which resulted in the fuel valve inadvertently being shut off and the subsequent total loss of engine power. Contributing causes included the lack of a flight instructor in the ground control station, inadequate maintenance procedures, and the operator's inadequate surveillance of the RPA program. The board also issued 22 safety recommendations and expressed concern about the practice of operating remotely piloted vehicles in the National Airspace System. Areas of particular concern included the design and certification of the uncrewed aircraft system, pilot qualification and training, the integration of remotely piloted aircraft into the air traffic management system, and the lack of audio records of RPA operations-related communications. NTSB Chairman Mark Rosenker said the recommendations addressed "a wide range of safety issues involving the civilian use of unmanned aircraft."[7]

---

7. Mary Grady, "NTSB Cites Range of Safety Issues in UAV Investigation," AV Flash News, AVweb, October 17, 2007, *http://www.avweb.com/avwebflash/news/NTSB_CustomsBorderPatrol_UAVcrash_196405-1.html*, accessed June 20, 2011.

"This investigation has raised questions about the different standards for manned and unmanned aircraft and the safety implications of this discrepancy,"[8] said Rosenker. He noted that the pilot's console had been known from past experience to lock up frequently, but this issue was not resolved. In fact, a logbook in the ground control station revealed 16 lockups in the 4½ months prior to the accident. No data had been recorded to explain these deficiencies. "Such conditions would never be tolerated in the cockpit of a manned aircraft."[9]

In the course of the investigation, the board determined that the CBP had been unable to certify the airworthiness of the MQ-9. Due to national security issues and past experience with similar vehicles, the FAA temporarily waived this requirement for the issuance of a Certificate of Authorization to operate in the national airspace along with all other civil and commercial air traffic.

Well-funded maintenance programs and adequate spare-parts support are essential to safe flight operations. Prior to the accident, the most recent maintenance performed on the MQ-9 had been a scheduled 200-hour inspection on April 21, 2006. General Atomics and CBP personnel told investigators that, due to limited funding, there was virtually no stock of spare parts available at Libby Army Airfield. Any necessary components were ordered, as needed, from the manufacturer.[10]

After the MQ-9 accident, maintenance personnel admitted that parts swapping had become commonplace because very few spare parts had been purchased with the vehicle. When the avionics technician swapped the main processor cards between the PPO-1 and PPO-2 consoles prior to the accident, it appeared to mitigate the immediate problem, but it was not good practice. Neither the operating agency—the CBP—nor its contractors had specified minimum essential equipment required for safe flight, another fundamental safety requirement in crewed aviation. Additionally, there was no documented program specifying how maintenance tasks were to be performed, especially after repeated similar anomalies. No formal procedures stipulated how, when, or by whom aircraft maintenance was to be inspected or released for return to flight after maintenance. This situation was a classic example of the normalization of deviance in which bad practices were allowed to continue simply because there had yet to be a bad result.[11]

---

8. Ibid.
9. Ibid.
10. "NTSB Accident Brief: General Atomic Predator-B, April 25, 2006, Nogales, AZ (Docket CHI06MA121)."
11. Ira J. Rimson and Ludwig Benner, Jr., "The Sky Isn't Falling—Or Is It? Part 2," *Journal of System Safety* 44, no. 4 (July–Aug. 2008).

The accident highlighted problems with system prioritization. The MQ-9 was operated by means of C-band signal communications, which provided for line-of-sight control. A secondary means of control was provided through the Iridium satellite communication system. However, when using the Iridium system, the vehicle could be controlled solely by means of autopilot hold modes, all of which (altitude, airspeed, and heading) must be active for the Iridium satellite to control the aircraft. The hold modes were activated prior to loss of uplink. If both the satellite and line-of-sight (LOS) uplinks are active, the LOS link has priority and Iridium data are ignored. When the fuel was cut off to the engine, the aircraft began shutting off electrical equipment to conserve battery power. The Iridium system was one of these items, and the aircraft's auto-ignition system will not work unless the system is operable.[12]

As the MQ-9 continued to automatically shut down what should have been noncritical systems, the transponder was deactivated. The transponder is vitally important to air traffic control because it provides an enhanced electronic signature, identification code, and altitude information that are presented on the controller's radar display. Without an operational transponder, controllers lost secondary radar contact with the MQ-9 as well as altitude information. Unaware that the MQ-9 was about to crash, controllers expected the aircraft to fly the same course it had during several previous lost-link events—through a corridor northeast of Nogales at 15,000 feet toward Libby Army Airfield. Following loss of radar contact, the RPA pilot informed controllers that he did not know the location of the aircraft. Although FAA air traffic controllers considered the loss of radar contact and radio communications with the MQ-9 to be an emergency, neither the pilot nor the controllers declared an emergency.

Investigators also examined the switch configuration on the PPO consoles. In the pilot configuration, the control lever was used to control both the fuel valve and propeller feather servo. This lever had a linear analog range running from zero to 100 percent, divided into thirds for normal operation, engine shutdown, and propeller-feather/engine-shutdown. A shutdown engine could be restarted if the switch was positioned in the normal range and engine controls were in auto mode. The system was programmed to attempt automatic restart if propeller speed fell below 1,488 rpm and the condition lever was in the normal run range. The engine could not be restarted with the lever in the shutdown range because this closed the fuel valve, shutting off fuel flow to the engine.

The investigation board learned that there was an audible warning when engine failure occurred. Confusingly, the same tone was also used for every

---

12. Ibid.

system warning; there was no distinctive tone for loss of engine power. The avionics technician heard the tone but believed it signified the loss of the satellite link. The pilot may also have misinterpreted the aural warning but should have noticed the loss of engine torque and an exhaust gas warning that were displayed on his control console.[13]

David Gasho, director of CBP air operations at Libby Field, was more critical of the drone pilot's actions. "This was one of these instances where he would have been better off not touching it," said Gasho. "He just panicked, hit the button and threw away a $7-million airplane."[14]

According to some safety analysts, it was apparent from the NTSB investigation report that neither the vehicle's manufacturer nor its operator nor the FAA gave credible consideration to the transition from the aircraft's military applications to its use as a civilian workhorse. Military RPA operate in aeronautical environments (i.e., restricted special-use airspace) sparsely populated by other aircraft. In civilian roles, they must operate in, or adjacent to, areas dense with air civil and commercial traffic.[15]

As is frequently the case, this mishap might have been easily avoided. A simple switch lockout could have prevented the pilot from inadvertently shifting aircraft control from one console to the other until the condition lever positions matched. Such a device—used effectively in everything from automobiles to nuclear weapons—would have provided a final opportunity to avoid disaster.[16]

Human Factors Analysis and Classification System

Anthony Tvaryanas, William Thompson, and Stefan Constable created the Human Factors Analysis and Classification System to provide a quantitative analysis of the role and patterns of active and latent human failures in RPA mishaps using standardized human-factors taxonomy. The HFACS is a model of accident causation based on the premise that latent failures associated with organizational influences, unsafe supervision, and unsafe preconditions lead to active failures (unsafe acts). In this model, operator error is the result of a series of preexisting latent failures. Human-factors analysts consider targeting these

---

13. "NTSB Accident Brief: General Atomic Predator-B, April 25, 2006, Nogales, AZ (Docket CHI06MA121)."

14. Brian Bennett, "Predator drones take on stateside fire, flood patrol," *Los Angeles Times*, Sept. 13, 2011.

15. Rimson and Benner, "The Sky Isn't Falling—Or Is It? Part 2."

16. Ibid.

latent failures a more effective approach to mishap prevention than simply reacting to specific acts by operators.[17]

The study by Tvaryanas, Thompson, and Constable provided a quantitative analysis of the role of both active and latent human failures in RPA mishaps within the U.S. military services. This was significant because latent failures more frequently contribute to mishaps than do active failures and because such a comparison might determine which human factors are common and likely inherent in all RPA operations and which are service specific (reflecting different policies and processes) or are unique to specific models of aircraft.

The researchers examined 271 mishaps spanning a period between fiscal years 1994 and 2003 that involved a variety of remotely piloted aircraft used by the Air Force, Army, and Navy/Marines. To reduce the heterogeneity of data between the services, all mishap reports pertaining to uncrewed target drones were excluded. The remaining 221 mishaps were submitted to initial analysis using the HFACS taxonomy. Typically, any single mishap had several contributing human factors associated with it. The investigators used only those causes and contributing factors identified during the original accident investigations. Although no new causal factors were identified, mishap codes reflected any instance in which a reasonable inference could be made as to embedded human causal factors based on the original mishap narrative, findings, or recommendations.[18]

Human-factors mishaps were found to be most prevalent in the Air Force, followed by the Navy/Marines, and finally the Army. Researchers found that the pattern of latent failures predisposing error among RPA operators varied markedly between the services, however. Latent failures in Air Force RPA mishaps involved instrumentation and sensory feedback systems, automation, and channelized attention. Navy/Marine RPA mishaps were more often found to be associated with workload, attention problems, and risk-management issues. Latent failures associated with Army RPA operators typically involved procedural guidance and technical publications, training, operator overconfidence, communications, and crew coordination. Skill-based errors were more common among Air Force operators while Army operators were more prone to procedural violations. There was no difference among the services with regard to the frequency of decision errors.[19]

Tvaryanas, Thompson, and Constable made several service-specific recommendations based on the results of their study. First, they recommended

---

17. Tvaryanas, Thompson, and Constable, "U.S. Military Unmanned Aerial Vehicle Mishaps."

18. Ibid.

19. Ibid.

that the Air Force undertake a comprehensive program to evaluate and optimize RPA operator selection and training as well as human-machine interface (ground control station) design. Second, they suggested that the Army improve technical publications and checklists and emphasize crew resource management during initial RPA operator training. Third, they recommended that the Navy/Marines improve RPA operator task and workstation design, assess manpower requirements, develop empirically based training programs and formal procedures, and institutionalize operational risk management at all levels of RPA operations. Since many electromechanical malfunctions were deemed to be manifestations of latent failures in acquisition processes, they also recommended that the Department of Defense refocus the investigational spotlight from immediate mechanical failures as the cause of RPA mishaps to failures in the organizational culture, management, or structure of such processes.[20]

Evidence-Based Prioritization of Human-Factors Issues

Analysis of postaccident data remains the predominant means of investigating the causal role of human error in RPA mishaps. Studies by researchers at the U.S. Army Aeromedical Research Laboratory indicate that the HFACS provides the most useful and detailed information for examining individual human errors. [21]

The HFACS is based on James Reason's "Swiss cheese" model of system mishaps, which incorporates not only the active errors that lead to a mishap, but also the latent errors. Reason described these latent conditions as "holes" in the layers of defense ("slices of cheese") established to guard against such error. These defenses include standard operating procedures, supervisory practices, managerial decisions, and corporate leadership. These "holes" of vulnerability and risk are not static. Over time, they change size and location within a particular layer of defense. In fact, they are constantly changing size, shape, and location as an organization or project evolves over time in ways related to changing circumstances. When by chance the holes align, a mishap trajectory (the path from hazard to mishap) occurs.[22]

---

20. Ibid.

21. P.A. Leduc, C.E. Rash, and M.S. Manning, "Human Factors in UAV Accidents," *Special Operations Technology* 3, no. 8 (2005).

22. Scott A. Shappell and Douglas A. Wiegmann, "The Human Factors Analysis and Classification System (HFACS)," FAA report no. DOT/FAA/AM-00/7 (2000).

Using the HFACS as a primary tool to investigate and document an individual mishap results in a case study of system failure. As a secondary investigative tool, the HFACS can be used to evaluate a collection of mishap reports and create a catalog of observed frequency or prevalence of active and latent failures. Studies show that mishaps tend to fall into recurrent patterns in which the same set of circumstances provokes similar errors regardless of the individuals involved. By identifying recurring patterns, human-factors analysts can develop more systematic approaches to mishap prevention.[23]

Applying the HFACS to the MQ-9 Mishap

Geoff Carrigan, Dave Long, M.L. Cummings, and John Duffner used the HFACS approach to examine a variety of latent issues preceding the loss of the MQ-9, and they found significant organizational influences contributing to the crash. As with any accident, the Nogales mishap resulted from the convergence of a variety of factors rather than a single event.

In the immediate circumstances leading up to the accident, the first unsafe precondition arose shortly after takeoff, when the pilot failed to increase the altitude setting for the lost-link flight profile. He was apparently unaware that the profile had been preset for a lower flight pattern than the approved altitude and that the setting needed to be updated. This was a precondition for unsafe acts because the pilot did not fully understand the lost-link profile. Additionally, a higher altitude setting may have provided additional time to reestablish communications with the RPA if the uplink was still functional.[24]

A second precondition for unsafe acts was established when the telemetry data locked up, forcing the pilot to operate without current information on the state of the aircraft. The lockup issue was a recurring problem that had not been addressed because the operator relied on the built-in lost-link flight plan. When he lost telemetry due to monitor lockup, the pilot followed the common practice of switching control to the second console but failed to follow the checklist while doing so. Checklist procedures stressed the importance of making sure that both consoles were set to the same control positions before switching control of the aircraft from one to the other. Failure to follow

---

23. Anthony P. Tvaryanas and William T. Thompson, "Unmanned Aircraft System Operator Error Mishaps: An Evidence-Based Prioritization of Human Factors Issues" RTO-MP-HFM-135, *http://www.wpafb.af.mil/shared/media/document/AFD-090417-032.pdf*, accessed June 10, 2011.

24. Geoff Carrigan, Dave Long, M.L. Cummings, and John Duffner, "Human Factors Analysis of a Predator B Crash," from the Unmanned Systems North America Conference, San Diego, CA, June, 2008.

the checklist was a significant unsafe act leading to the crash. The pilot's lack of awareness of the mismatched control settings led to mode confusion because the system behaved differently from the user's expectations.

Another unsafe act occurred when the payload operator left the ground control station during the switchover. According to the checklist, both operators were required to remain in their seats during transfer of control in case it became necessary to revert controls. Additionally, during control switchover, the avionics technician should have assumed the duties of copilot to help perform checklist items. Failure to do so was another unsafe act. If the flightcrew had followed checklist procedures, the switch-setting error may have been avoided.[25]

The design of the backup communication mode set up another unsafe precondition for unsafe acts. Control via satellite link was limited to preset autopilot hold modes, but line-of-sight communications automatically took priority when it was available. In the event of engine shutdown, low-priority electrical systems were shut down to conserve battery power. The satellite communication link was one of the first items shed during this process, making it impossible for the automatic engine restart system to function. Having the vehicle equipped with satellite link and auto-restart systems gave operators a false sense of security that was shattered when the satellite link was shut down. This single-point failure was a latent error that went unnoticed, leading to additional confusion because the crew believed the vehicle was following a lost-link flight plan (i.e., holding altitude and circling).[26]

Crew readiness also affected the preconditions for unsafe acts and unsafe supervision issues. The pilot, a 35-year-old employee of General Atomics, held both commercial and certified instructor pilot certificates, with single-engine land, multiengine land, and instrument ratings, along with an advanced ground instructor certificate. His FAA first-class medical certificate had no limitations. He had a total of 3,571 total flight hours, including 519 hours of Predator-A flight time and 27 hours of Predator-B flight time. The latter included nine flights with 5 hours of training. Since there were no Predator-B simulators available prior to the mishap, all flight training was completed using the actual vehicle and ground control station. At the time, CBP pilots were required to have 200 hours of flight time in crewed aircraft and 200 non-type-specific hours in RPA. Additionally, the CBP required that "all operators shall also be certified by the contractor as being fully capable of maintaining and operating the 'Predator B' UA and its associated equipment."[27]

---

25. Ibid.

26. Ibid.

27. "NTSB Accident Brief: General Atomic Predator-B, April 25, 2006, Nogales, AZ (Docket CHI06MA121)."

General Atomics used a training syllabus, approved by the Air Force, to train pilots to operate the CBP vehicle for the Air Force. Once a pilot completed the syllabus, General Atomics presented the completed training records, which had been approved by the Air Force Government flight representative, to CBP representatives, who either approved or disapproved the specific pilot to operate the vehicle. At the time of the accident, the CBP did not have a fully trained Government flight representative on its staff.

The mishap pilot's training was documented on several different Department of Defense and Air Force forms. One indicated that on February 17, 2006, the Air Force flight representative approved the pilot to begin MQ-9 training. The pilot completed that training on March 24, 2006. On May 5, after the date of this accident, the Air Force flight representative disapproved the pilot's request for approval to act as a MQ-9 pilot and cited that the pilot had not completed some training modules. In fact, 5 of 32 training events listed on the documents had not been completed during the pilot's training course, including Mission Planning/Briefing/Debriefing, Handover Procedures—Ground, Mission Monitor/Multifunction Workstation Procedures, Operational Mission Procedures, and Handover Procedures—Airborne. Training records provided to the NTSB listed no specific training on procedures to switch control of the MQ-9 from one PPO to the other.[28]

As noted, the pilot flying during the mishap had obtained most of his RPA flight experience in the MQ-1 Predator-A. Although both vehicles use a similar control console, there is one significant difference. When switching control between PPO consoles while flying the Predator-A, the condition levers do not have to be synchronized. Stress and mode confusion at the start of the mishap sequence may have contributed to the pilot's inadvertent use of Predator-A procedures for the Predator-B.

Inadequate supervision was a factor during the pilot's training and certification process. According to the training syllabus, Air Force grading criteria are used to evaluate the pilot's performance during training. According to CBP, General Atomics contacted one of the CBP personnel being trained as a Government flight representative and requested that the mishap pilot be added to the approved list of CBP pilots prior to Air Force approval, knowing that the pilot had not completed his training modules. The trainee reportedly gave General Atomics a verbal approval so that the pilot could operate the CBP vehicle—but only when an instructor pilot was physically present in the ground control station. According to NTSB investigators, this verbal approval was not standard practice for CBP. This act was a supervisory violation because

---

28. Ibid.

allowing a pilot who had not yet completed the full training syllabus to operate the MQ-9 significantly increased the probability of operational errors that might not have been committed by a fully trained pilot.[29]

Contributing factors at the organizational level included failed training oversight and complacency with regard to vehicle readiness. Without an available simulator, MQ-9 pilots could not practice emergency scenarios. This lack of training contributed to inaccurate mental models of the vehicle's limits and corresponded directly to the pilot's lack of personal readiness. Failure to address the issue of repeated display lockups—a total of 16 incidents over the preceding 5 months—and lack of sufficient spare parts suggest a lack of organizational oversight of the agency's RPA maintenance program.

By applying the HFACS and other methods to the Nogales MQ-9 mishap, three general problem areas can be identified that apply to all RPA systems. First, human factors should be considered when designing remotely piloted aircraft and their ground control interfaces. Systems should be designed to prevent the likelihood of mode confusion. Improved warning and caution systems, additional automation, and switch lockouts could prevent similar mishaps in the future. Second, standard analytical models for RPA systems reliability could be used to identify potential single-point failures and other hazards. Finally, improved organizational oversight and provision of adequate maintenance support can enhance safety.

The NTSB investigation identified many of the same causes as the HFACS analysis. While the NTSB focused on immediate causes of the mishap and on human errors committed by the operators, the HFACS analysis identified these errors as byproducts of improper design and poor management. In mishaps involving significant human-factors issues, the HFACS can be used to determine why the human error occurred by identifying organizational and latent errors and can point toward how such problems might be mitigated in the future.[30]

Lessons Learned

Standard NTSB investigative methods coupled with HFACS analysis yields numerous lessons from this single mishap.

- Confusing control configurations should be eliminated during the design phase. Human-machine interfaces should be designed to prevent the likelihood of mode confusion.

---

29. Carrigan et al., "Human Factors Analysis of a Predator B Crash."
30. Ibid.

- To avoid confusion, distinctive tones should be provided for different warnings.
- Switch lockouts should be used to prevent inadvertently moving critical controls.
- Two qualified pilots should be present in the GCS prior to switching control of the aircraft from one console to the other.
- Always follow the procedures outlined in the checklist. Failure to follow the checklist can lead to missing critical steps.
- Ensure the proper settings of switches on control panels in order to avoid loss of mode awareness.
- Maintenance and training standards should be identical for crewed aircraft and RPA.
- Battery power conservation systems should leave mission critical systems operating when shutting down noncritical systems.
- The pilot should declare an emergency in a timely fashion to allow for a rapid contingency response.
- Latent organizational influences (inadequate training, supervision, etc.) can lead to normalization of deviance and set the stage for mishaps.
- Human factors should be considered when designing remotely piloted aircraft and their ground control interfaces.
- Standard analytical models for RPA systems reliability could be used to identify potential single-point failures and other hazards.
- Improved organizational oversight and provision of adequate maintenance support can enhance safety. Well-funded maintenance programs and adequate spares support are essential to safe flight operations.

Conclusion
The Lessons of History

By now, the reader should have a grasp of the various factors affecting aircraft safety. Some of these are unique to RPA while most are common to all aircraft operations. Inadequate design (particularly with regard to the human-machine interface) and organizational deficiencies are frequent contributors to mishaps despite more than a century of experience with powered flight. The expanding role of RPA for civil and military applications, and their imminent introduction into the National Airspace System, makes it imperative to address relevant safety issues as soon as possible. RPA safety has a direct impact on acquisition, operations, logistics, and regulatory issues. According to two reports by the Office of the Secretary of Defense, the reliability and sustainability of RPA underlies their affordability, mission availability, and their acceptance into civil airspace.[1] As with any rapidly evolving technology, RPA development has not always progressed smoothly. Mishap rates for RPA have been high and deserve additional study in order to benefit from the lessons of history.

Learning from past experience is fundamental to the development of safe and efficient new systems and to improving existing systems. Future mishaps might be most easily avoided through collecting, archiving, and studying past accidents to learn valuable lessons. John Langford, Chairman and President of Aurora Flight Sciences Corporation, observed of RPA development, "There are many issues, and one of the fundamental ones is that in robotic systems, in an important way, you capture all of the lessons of all the previous mistakes and incidents that have gone before you."[2]

But are these lessons truly being learned? Many of the preceding case studies illustrate patterns of similar failures or contributing factors. NASA's DAST program suffered multiple mishaps, each attributed to a combination of design, operational, and management factors. Although there were dissenting opinions regarding the findings in both cases, it may be hoped that the ensuing debate had a positive affect on the programs that followed. Aurora Flight Sciences now

---

1. Tvaryanas and Thompson, "Unmanned Aircraft System Operator Error Mishaps: An Evidence-based Prioritization of Human Factors Issues."
2. Rimson and Benner, "The Sky Isn't Falling—Or Is It? Part 2."

specializes in the design and construction of uncrewed aircraft, but prior to joining the ERAST project in 1993, the company had built only one remotely piloted proof-of-concept demonstrator. While testing several advanced proto-types in the span of just a few years, Aurora encountered a steep learning curve in the field of RPA and a prodigious mishap rate. Some losses are expected during the development of cutting-edge technology, but Aurora's early losses resulted from failures of proven hardware, insufficient training, and deficiencies in the design of the human-machine interface. This latter factor highlights the need to eliminate design deficiencies—a frequent cause of aircraft mishaps—as early as possible. Potential design problems should be addressed before the aircraft or GCS configuration is finalized for production.

One modern approach to aircraft testing and acquisition, known as rapid prototyping, can inadvertently introduce design problems. General Atomics employed rapid prototyping methods to develop a variety of successful RPA in the MQ-1 Predator and MQ-9 Reaper design families, allowing the company to build fully functional systems for demonstration to potential customers. An urgent need for such systems to support military operations in Southwest Asia pushed these systems into operational use before their designs were fully refined. According to one senior intelligence official, "We were producing Predators and Reapers as fast as we could, but the company was maxed out."[3] As a result of quickly moving from rapid-prototyping to fielding an operational system, some important developmental steps may have been omitted in developing the human-machine interface.

Mark Pestana, describing some of the challenging aspects of the MQ-9 control console, noted that many of the "switches" are actually menu-driven keyboard commands, embedded in software and displayed on four screens. One displays a forward-facing camera view; another displays maps with a moving aircraft icon; the remaining two display systems information, cautions and warnings, and operations menus selectable through a keyboard and mouse or trackball. Only one page of systems/cautionary information is shown at any time, but more than 60 additional pages are available.[4]

"Often," according to Pestana, "my attention is diverted from the primary flight displays (the forward screen camera view is overlaid with 'head up' flight information) because I'm 'head down' using a keyboard to access the information screen displays and activate 'switches,' or I'm moving the trackball cursor on the upper map screen to access menus for changing radio frequencies." With this display format, incorporation of timely software feedback and

---

3. John A. Tirpak, "Lifesaving Liberty," *Air Force Magazine* 94, no. 4 (April 2011).

4. Pestana, "Flying NASA Unmanned Aircraft," p. 5.

user-friendly displays are absolutely imperative. "Embedded menus can add to the time required to accomplish time-critical actions, and divert attention from primary piloting tasks."[5]

It is also important to remember that although there are no humans on board remotely piloted aircraft, there are numerous humans involved in all aspects of RPA operations. Human factors affect RPA safety at every level of design, management, maintenance, and flight operations. The case studies cited in this volume contain a shocking number of incidents involving improper procedures, training deficiencies, management problems, and supervisory failures. Additionally, it seems apparent that many accidents could be prevented with simple modifications to the human-machine interface. For example, inexpensive video cameras could provide an RPA operator with peripheral vision. A ground cockpit could be equipped with an aural cue to alert the pilot to engine failure. A switch lockout or onscreen popup display could serve as a fail-safe to prevent mistakes when switching from one control console to another. Such common sense ideas seem fundamental, but as of this writing, none of these concepts have been adopted.

Because human factors are consistently cited as a major cause of RPA accidents, an understanding of the causal factors associated with these accidents is essential for improving the reliability of remotely piloted aircraft. This is especially true if RPA are to be integrated into the National Airspace System. Review and analysis of accident data can help researchers identify important human-factors issues, but different operating agencies use different investigation and reporting taxonomies. The Human Factors Analysis and Classification System offers investigators a standardized human-factors taxonomy focusing on the root causes of most RPA accidents.[6]

Culturally, most organizations would prefer not to publicize their failures, but sometimes it is necessary for the greater good. That this volume contains case studies collected from numerous Government and industry sources clearly illustrates that no organization in this field is immune to failure. Hopefully, the lessons learned from these studies will prevent future mishaps and promote safe practices in the use of remotely piloted aircraft and, indeed, any type of aircraft.

---

5. Ibid., p. 6.

6. Williams, "A Summary of Unmanned Aircraft Accident/Incident Data: Human Factors Implications."

Bibliography

Reports, Papers, Articles, and Presentations

Arnaiz, Henry H., and Paul C. Loschke. "Current Overview of the Joint NASA/USAF HiMAT Program." NASA CP-2162, 1980.

Asim, Muhammad, Nadeem Ehsan, and Khalid Rafique. "Probable Causal Factors in UAV Accidents Based on Human Factor Analysis and Classification System." Presented at the 27th Congress of International Council of the Aeronautical Sciences, September 19–24, 2010, Nice, France.

Ault, Stanley K., and Sarita S. May. "RAPTOR Demonstrator Aviation Safety Document, Revision 3." Lawrence Livermore National Laboratory, Livermore, CA (1994).

Barnes, M.J., B.G. Knapp, B.W. Tillman, B.A. Walters, and D. Velicki. "Crew Systems Analysis of Unmanned Aerial Vehicle (UAV) Future Job and Tasking Environments." Army Research Laboratory, Aberdeen Proving Ground, ARL-TR-2081, 2000.

Barnes M.J., and M.F. Matz. "Crew Simulations for Unmanned Aerial Vehicle (UAV) Applications: Sustained Effects, Shift Factors, Interface Issues, and Crew Size." Presented at the Human Factors and Ergonomics Society 42nd Annual Meeting, Chicago, IL, 1998.

Bauer, Jeffrey E., and Edward Teets. "An Impact-Location Estimation Algorithm for Subsonic Uninhabited Aircraft." NASA TM-97-206299, 1997.

Biggerstaff, S., D.J. Blower, C.A. Portman, and A.D. Chapman. "The Development and Initial Validation of the Unmanned Aerial Vehicle (UAV) External Pilot Selection System." Naval Aerospace Medical Research Laboratory (NAMRL) 1398, 1998.

Brown, L.E. Jr., M.H. Roe, and R.A. Quam. "HiMAT Systems Development Results and Projections." Society of Automotive Engineers (SAE) Aerospace Congress and Exposition, SAE Paper 801175, 1980.

Brown, L.E., M. Roe, and C.D. Wiler. "The HiMAT RPRV System." AIAA Aircraft Systems and Technology Conference, AIAA Paper 78-1457, 1978.

Calhoun, G.L., M.H. Draper, H.A. Ruff, and J.V. Fontejon. "Utility of a Tactile Display for Cueing Faults." Presented at the Human Factors and Ergonomics Society 46th Annual Meeting, Santa Monica, CA, 2002.

Carrigan, Geoff, Dave Long, M.L. Cummings, and John Duffner. "Human Factors Analysis of a Predator B Crash." Presented at the AUVSI Unmanned Systems North America, San Diego, CA, 2008.

Colella, Nicholas J., and Gordon S. Wenneker. "Pathfinder and the Development of Solar Rechargeable Aircraft." *Energy & Technology Review* (July 1994).

Cross, Ken. "UAV Mishap Summary Report." AS&M System Safety, NASA DFRC, 2003.

Deets, Dwain A., and John W. Edwards. "A Remotely Augmented Vehicle Approach to Flight Testing RPV Control Systems." NASA TM-X-56029, 1974.

Del Frate, John H. "Four Remotely Piloted Aircraft Mishaps—Some Lessons Learned." Presented at the Association for Unmanned Vehicles Systems International (AUVSI) conference, 1996.

Del Frate John H., and Gary B. Cosentino. "Recent Flight Test Experience with Uninhabited Aerial Vehicles at the NASA Dryden Flight Research Center." NASA TM-1998-206546, 1998.

Draper, M., G. Calhoun, H. Ruff, D. Williamson, and T. Barry. "Manual Versus Speech Input for the Unmanned Aerial Vehicle Control Station Operations." Presented at the Human Factors and Ergonomics Society 47th Annual Meeting, 2003.

Gopher, D., M. Weil, and T. Bareket. "Transfer of Skill from a Computer Game to Flight." *Human Factors* 36, no. 3 (1994).

Duke, E.L., F.P. Jones, and R.B. Roncoli. "Development of a Flight Test Maneuver Autopilot for a Highly Maneuverable Aircraft." AIAA Paper 83-0061, 1983.

Edwards, John W. "Flight Test Results of an Active Flutter Suppression System Installed on a Remotely Piloted Vehicle." NASA TM-83132, 1981.

Grose, David L. "The Development of the DAST I Remotely Piloted Research Vehicle for Flight Testing an Active Flutter Suppression Control System." NASA CR-144881, 1979.

Hall, E.M., and W.C. Tirre. "USAF Air Vehicle Operator Training Requirements Study." Air Force Research Laboratory, AFRL-HE-BR-SR-1998-0001, 1998.

Hutchison, Matthew G., Matthew T. Velazquez, and David W. Vos. "Flight Testing of the Theseus Prototype." Presented at the AUVSI Annual Symposium and Exhibition, 1997.

Israel, Kenneth R. "UAV Annual Report: FY 1996." Defense Airborne Reconnaissance Office, 1996.

Johnson, Chris W., and Christine Shea. "The Hidden Human Factors in Unmanned Aerial Vehicles." In *Proceedings of the 26th International Conference on Systems Safety Proceedings of the 26th International Conference on Systems Safety*. Unionville, VA: International Systems Safety Society, 2008.

Kempel, Robert W., and Michael R. Earls. "Flight Control Systems Development and Flight Test Experience with the HiMAT Research Vehicles." NASA TP-2822, 1988.

Manning, Sharon D., C.E. Rash, P.A. LeDuc, R.L. Noback, and J. McKeon. "The Role of Human Causal Factors in U.S. Army Unmanned Aerial Vehicle Accidents." U.S. Army Aeromedical Research Laboratory (USAARL) Report No. 2004-11, 2004.

McCarley, Jason S., and Christopher D. Wickens. "Human Factors Concerns in UAV Flight." Institute of Aviation, Aviation Human Factors Division, University of Illinois, Technical Report AHFD-05-05/FAA-05-01, 2005.

Molloy, R., and R. Parasuraman. "Monitoring an Automated System for a Single Failure: Vigilance and Task Complexity Effects." *Human Factors* 38, no. 2 (1996).

Morris, George A. "Hazards Analysis of the RAPTOR Project Flights." Lawrence Livermore National Laboratory, 1993.

Murrow, H.N., and C.V. Eckstrom. "Drones for Aerodynamic and Structural Testing (DAST)—A Status Report." Presented at the AIAA Aircraft Systems and Technology Conference, Los Angeles, CA, August 21–23, 1978.

Myers, Albert F. "Engineering Analysis Results of HiMAT Flight Mishap, July 8, 1980 (H1-5-10)." NASA Report E-EDC(AFM)3397, 1980.

Nas, Michael. "The Changing Face of the Interface: An Overview of UAS Control Issues and Controller Certification." Unmanned Aircraft Technology Applications Research Working Group 27 (UATAR WG-27), February 20, 2008.

Parasuraman, R., and V. Riley. "Humans and Automation: Use, Misuse, Disuse, Abuse." *Human Factors* 39, no. 2 (1997).

Reed, R. Dale. "Flight Research Techniques Utilizing Remotely Piloted Research Vehicles." Technical Report (TR) AGARD-LS-108, Paper No. 8, Research Engineering, NASA Dryden Flight Research Center (1980).

Reed, R. Dale. "RPRVs—The First and Future Flights." *Astronautics and Aeronautics* (April 1974).

Rezek, Terrence W. "Unmanned Vehicle Systems Experiences at the Dryden Flight Research Facility." NASA TM-84913, 1983.

Rimson, Ira J., and Ludwig Benner, Jr. "The Sky Isn't Falling—Or Is It? Part 1." *Journal of System Safety* 44, no. 2 (March–April 2008).

Rimson, Ira J., and Ludwig Benner, Jr. "The Sky Isn't Falling—Or Is It? Part 2." *Journal of System Safety* 44, no. 4 (July–August 2008).

Rimson, Ira J., and Ludwig Benner, Jr. "The Sky Isn't Falling—Or Is It? Part 3." *Journal of System Safety* 44, no. 6 (November–December 2008).

Rutley, Mark. "On Target—Human Factors: UAV's." Canadian Forces Environmental Medical Establishment, 2010.

Sackett, Steven J., et al. "Observations and Analysis of the LLNL Technical Advisory Team Supporting the Investigation of the RAPTOR Demonstrator Crash." Lawrence Livermore National Laboratory, 1994.

Schmidt, J., and R. Parker. "Development of a UAV Mishap Human Factors Database." Presented at the 1995 AUVSI conference.

Schreiber, B.T., D.R. Lyon, E.L. Martin, and H.A. Confer. "Impact of Prior Flight Experience on Learning Predator UAV Operator Skills." Air Force Research Laboratory, AFRL-HE-AZ-TR-2002-0026, 2002.

Shane, Douglas B. "One Look at the Role of the Test Pilot in UAV Development." Presented at the Society of Experimental Test Pilots (SETP) 40th Symposium, 1996.

Shappell, Scott A., and Douglas A. Wiegmann. "A Human Error Approach to Accident Investigation: The Taxonomy of Unsafe Operations." *International Journal of Aviation Psychology* 7, no. 4 (1997).

Shappell, Scott A., and Douglas A. Wiegmann. "The Human Factors Analysis and Classification System (HFACS)." FAA report number DOT/FAA/AM-00/7, 2000.

Sherry, Richard R., and Frank E. Ritter. "Dynamic Task Allocation: Issues for Implementing Adaptive Intelligent Automation." School of Information Sciences and Technology, Pennsylvania State University, Technical Report No. ACS 2002-2, 2002.

Spravka, John J., Deborah A. Moisio, and Mary G. Payton. "Unmanned Air Vehicles: A New Age in Human Factors Evaluations." *Flight Test—Sharing Knowledge and Experience*, Meeting Proceedings RTO-MP-SCI-162, 2005.

Tirpak, John A. "Lifesaving Liberty." *Air Force Magazine* 94, no. 4 (April 2011).

Tvaryanas, Anthony P. "Visual Scan Patterns During Simulated Control of an Uninhabited Aerial Vehicle." *Aviation, Space, and Environmental Medicine* 75, no. 6 (2004).

Tvaryanas, Anthony P., William T. Thompson, and Stephan H. Constable. "U.S. Military Unmanned Aerial Vehicle Mishaps: Assessment of the Role of Human factors Using Human Factors Analysis and Classification

System." 311th Performance Enhancement Directorate, HSW-PE-BR-TR-2005-0001, 2005.

Walker, Laurence A. "Flight Testing the X-36—The Test Pilot's Perspective." NASA CR-198058, 1997.

Weeks, J.L. "Unmanned Aerial Vehicle Operator Qualifications." Air Force Research Laboratory, AFRL-HE-AZ-TR-2000-0002, 2000.

Williams, Kevin W. "A Summary of Unmanned Aircraft Accident/Incident DATA: Human Factors Implications." Federal Aviation Administration Report Number DOT/FAA/AM-04/24, 2004.

Weiss, K.A., N. Leveson, K. Lundqvist, N. Farid, and M. Stringfellow. "An Analysis of Causation in Aerospace Accidents." Presented at the 20th Institute of Electrical and Electronics Engineers (IEEE) Digital Avionics Systems Conference, 2001.

Wilson, J.R. "UAVs and the Human Factor." *Aerospace America* (July 2002).

X-34A Mishap Investigation Board. "Report of Findings: X-43A Mishap." NASA, 2002.

Books and Monographs

Dismukes, R.K., B.A. Berman, and L.D. Loukopoulos. *The Limits of Expertise: Rethinking Pilot Error and the Causes of Airline Accidents*. Burlington, VT: Ashgate Publishing, Ltd., 2007.

Hallion, Richard P., and Michael H. Gorn. *On the Frontier: Experimental Flight at NASA Dryden*. Washington, DC: Smithsonian Books, 2003.

Helmreich, R.L., and Ashleigh C. Merritt. *Safety and Error Management: The Role of Crew Resource Management*. Austin, TX: University of Texas, 2000.

Merlin, Peter W. *Ikhana Unmanned Aircraft System Western States Fire Missions*. Washington, DC: NASA SP-2009-4544, 2009.

Miller, Jay. *The X-Planes—X-1 to X-45*. Hinckley, UK: Midland Publishing, 2001.

Peebles, Curtis. *Road to Mach 10: Lessons Learned from the X-43A Flight Research Program.* Reston, VA: American Institute of Aeronautics and Astronautics, 2008.

Reed, R. Dale, with Darlene Lister. *Wingless Flight: The Lifting Body Story.* Washington, DC: NASA SP-4220, 1997.

Wiegmann, D.A., and S.A. Shappell. *A Human Error Approach to Aviation Accident Analysis, the Human Factors Analysis and Classification System.* Burlington, VT: Ashgate Publishing, Ltd., 2003.

About the Author

Peter W. Merlin is an aerospace historian with extensive knowledge of the various factors involved in aerospace mishaps. Under contract to the National Aeronautics and Space Administration (NASA) at Dryden Flight Research Center, Edwards, CA, since 1997, he has authored a variety of books, including several NASA Special Publications on aeronautical research projects. He served as coauthor of research pilot Donald Mallick's autobiography, *The Smell of Kerosene: A Test Pilot's Odyssey*, and *X-Plane Crashes: Exploring*  *Experimental, Rocket Plane, and Spycraft Incidents, Accidents and Crash Sites*, with Tony Moore. He has also authored several technical papers for the American Institute of Aeronautics and Astronautics, as well as numerous journal articles on aerospace history and technology. In addition, he serves as contributing editor for historical publications at Dryden and has appeared in more than a dozen documentary television programs for the Discovery Channel, the History Channel, National Geographic Channel, and others. He holds a bachelor of science degree in aviation management from Embry-Riddle Aeronautical University.

Index